Operations Management for Construction

Construction sites are challenging operations to run. From the setting up of the site to deciding the method of construction and the sequence of work and resourcing, construction managers find their skills and experience thoroughly tested.

This book explains the principles of operations management for construction, and how those principles work in practice. Procurement of materials, subcontractors and supply chain management are also carefully assessed, while explanations of contract planning, site organisation and work study provide further insights. With regulations increasingly impacting on the way sites are managed, relationships with third parties and the methods of successfully administering safety, quality and environment protection are spelt out.

Chris March has a wealth of practical experience in both the construction industry and teaching students. His down-to-earth approach and mixture of theory and real-life evidence from personal experience show just how to run a successful construction site operation.

Chris March is a graduate from Manchester University. He worked for John Laing Construction and later for John Laing Concrete where he became Factory Manager. On entering higher education he worked in both the UK and Hong Kong before joining the University of Salford becoming Senior Lecturer and then the Dean of the Faculty of the Environment. He is a former winner of the Council for Higher Education Construction Industry Partnership Award for Innovation.

Operations Management for Construction

Chris March

Spon Press
an imprint of Taylor & Francis
LONDON AND NEW YORK

First published 2009
by Taylor & Francis
2 Park Square, Milton Park, Abingdon, Oxon OX14 4RN

Simultaneously published in the USA and Canada
by Taylor & Francis
270 Madison Ave, New York, NY 10016

Taylor & Francis is an imprint of the Taylor & Francis Group, an informa business

Typeset in Sabon by
HWA Text and Data Management, London
Printed and bound in Great Britain by
TJ International Ltd, Padstow, Cornwall

British Library Cataloguing in Publication Data
A catalogue record for this book is available from the British Library

Library of Congress Cataloging-in-Publication Data
March, Chris.
Operations management for construction / Chris March.
p. cm.
Includes bibliographical references and index.
1. Building – Superintendence. 2. Business logistics. I. Title.
TH438.M307 2009
690.068´5–dc22 2008037922

ISBN10: 0-415-37112-0 (hbk)
ISBN10: 0-415-37113-9 (pbk)
ISBN10: 0-203-92803-2 (ebk)

ISBN13: 978–0–415–37112–4 (hbk)
ISBN13: 978–0–415–37113–1 (pbk)
ISBN13: 978–0–203–92803-5 (ebk)

Contents

Figures

Tables

Introduction

This book is one of three closely related texts, *Operations Management for Construction*, *Finance and Control for Construction* and *Business Organisation for Construction*; the reason for writing these books was the increasing awareness of the shortage of new texts covering the whole range of construction management. There are plenty of good recent texts appropriate for primarily final-year and postgraduate students, but they tend to be subject specific and assume a certain level of knowledge from the reader. It also means students find this cost prohibitive and tend to rely upon the library for access. (The research selectivity exercises have encouraged authors to write books based upon their research, for which credit has been given in the assessment, whereas none has been given to those writing textbooks.) The purpose of these three books is an attempt to give students the management vocabulary and understanding to derive greater value from these specialist texts.

The original intention was to write this with construction management undergraduate students in mind, but as the project developed it became clear that much of the subject matter was appropriate for all the construction disciplines. In more recent times the industry, being undersupplied with good construction graduates, has turned to recruiting non-cognate degree holders and many of these are and will study on Masters courses in construction management. These texts are ideally suited to them as background reading in giving a broad base of understanding about the industry.

Rather than having a large number of references and bibliographies at the end of each chapter, generally I have limited these to a few well-established texts, some referenced in more than one chapter, so the reader is directed to only a few if wishing to read further and in more depth on the subject. The chapters vary in length considerably depending upon the amount of information I believe is relevant at this level.

The aim of the book is to address the main issues associated with the site production activity, from setting up the site, planning the works, looking at productivity issues, the increasingly important issue of waste, managing suppliers and sub-contractors and total quality assurance.

Of the two related books, the first, *Finance and Control for Construction* is concerned with tracking through each stage of the process, the control of finance, with consideration taken of sustainability and the environment whilst *Business Organisation for Construction* is concerned with the running of the business. There is inevitably overlap in all three books, so I have cross-referenced from one book to another and within each subject, with the hope of aiding readers.

On a personal note I believe that there is no definitive way of managing and, as Mike Stoney, the Managing Director of Laing used to say to students, 'Don't copy me, it may not suit your personality, but watch and listen to other successful mangers and pinch the bits from them that suit you'. I totally agree and, for what it is worth, I have added some other comments and thoughts of other people that have influenced my way of thinking over the years.

My head master, Albert Sackett, who taught me to assume that everything was wrong until I could prove it correct. He would set an essay on say 'define the difference between wit and humour'. After he had marked it he would sit you in front of the class and then debate with you your answer. Having convinced you that he was right and you were wrong, would then reverse role and argue back the opposite way.

Godfrey Bradman, Chairman of Rosehaugh plc and now Chairman of Bradman Management services, reinforced my views from school of not simply accepting any thing you are told, but, in his case, also to have the ability to ask the right questions, usually simple ones such as 'why not?'

My father who taught me to accept failure was a fact of life and not to hide the fact, but to accept and admit it and get on with life having learnt from the experience. I also have found that by admitting it, 'the punishment' was always less than being found out. When in the precast factory it was always easier to advise the site manager that the load of components was going to be late or not delivered that day than await the angry phone call demanding to know what had happened. It also made sense because, although disappointed, they had more time to rearrange their own schedule of work.

Dorothy Lee, retired Deputy Director of Social Services in Hong Kong responsible for the Caritas operated Kai Tak East Vietnamese refugee camp who advised the small group I led in developing a self-build solution for refugees, which after many weeks of hard work was no longer required, said 'I know you will be disappointed, but remember you have grown a little more as a result.'

John Ridgway, explorer and outside activities course organiser, had at his School of Adventure, based in Sutherland, Scotland, the adage of positive thinking, self-reliance and to leave people and things better than you find them. He also made a very clear impression on me of the importance when in charge, to have the ability to stand outside the circle and view the problem from outside and allocate tasks without becoming too closely involved.

Don Stradling, the Personnel Director of Laing and senior negotiator with the Federation of Civil Engineers Employees with the very simple piece of advice that 'you should always keep the moral high ground'. How right he is. It is surprising the number of people that don't, and when confronted with one that does they almost invariable fail in the negotiation. It also results in having respect from those they have contact with, as they believe in your integrity and accept what you say is meant in an honourable way.

Finally Dennis Bate, member of the main board of Bovis Lend Lease who told me that he, throughout his life from leaving school at sixteen to become an apprentice joiner, strove to do whatever he did to the best of his ability and better than anybody else.

I wish to acknowledge the support and help given by so many in putting together these three books. From the construction industry, staff from Bovis Lend Lease, Laing O'Rourke and Interserve in particular, who have spent many hours discussing issues and giving advice. It was at Plymouth University that the idea to produce these texts was formulated and where colleagues gave me encouragement to commence, and then when at Coventry University, not only was this continued, but also doors were always open whenever I wished to consult on an idea or problem. My years at Salford, from 1987 onwards, when the we started the Construction Management degree, were of great significance in developing ideas on the needs of construction management students and this would not have been possible without the assistance and guidance from my colleagues there at the time, especially Tony Hills, John Hinks and Andy Turner, as well as the many supporting contractors always on hand to give advice, ideas and permit access to other colleagues in their organisations. There was a major contribution on the chapter on quality management by David Balkwell.

Permission to reproduce extracts from various British Standards is granted by the British Standards Institute.

Finally, to my wife Margaret who has to suffer many hours on her own whilst I locked myself away in the study, but never ceased to give her support and encouragement throughout.

Chris March

CHAPTER

Site organisation

1.1 Introduction

It is often said that if you get the blinding level correct at the onset of the construction of the building, then the rest of the building will be built correctly. So it is with site organisation. Get the basics in place and the contract stands a much better chance of being managed well. The problem is that often the time available, especially in traditional types of procurement, from the contract being awarded to the contractor being on site, there is only a limited time to think this process through properly. It is important that due attention is given to site organisation issues during the estimating process. In other forms of procurement, where the contractor or project manager are brought in during the design process, there is further opportunity to get these issues right as part of the continuous thinking process.

The prime areas of concern are those of:

- site boundaries
- access roads, on and off the site
- provision of services
- accommodation for contracting staff, sub-contractors and client's representatives
- material storage and handing
- waste disposal
- site logistics
- location of fixed plant
- hoardings
- communications
- security.

In all of the above, the most important document is the site plan, which shows existing services, the site boundary and the footprint of the building.

Other information needed includes elevation drawings, floor plans and other drawings which show the way the building is to be constructed, the quantities of bulk materials, component schedules, client's requirements such as staff accommodation and phased hand-overs, restrictions to work, maps showing roads to and around the site, and the programme for the contract. The majority, if not all, should be available in the preliminaries section of the contract documents (*Finance and Control for Construction,* Chapter 9). Finally a site visit is required to obtain a real understanding of the site and surrounding area, road congestion and so on.

1.2 Site boundaries

Where the site boundaries are and the position of existing buildings relative to the site of operations needs to be established. Any encroachment onto a third party's land can lead to dispute and extra costs. Since the ruling on Woollerton and Wilson *v* Richard Costain (1970), tower cranes overflying neighbours' properties can be considered an act of trespass unless the deeds permit it. If not, then permission has to be sought. It is not unusual for the owner of the affected land to request financial compensation in return. To avoid this and the delays and cost incurred in challenging the claim, alternative methods of construction might have to be considered, or the crane might have to be repositioned or changed to another type, such as one with a luffing jib.

The condition of buildings in close proximity to the works needs to be inspected and recorded before work commences, as contractors need to protect themselves against claims and only accept redress sought for further deterioration caused by the construction works. Equally, steps need to be taken to ensure that trees close to the border are not damaged, especially if they have a tree preservation order placed on them.

1.3 Access roads

Properly positioned and appropriately constructed access roads are essential to the successful running of the contract, as they are an integral part of the production process. They are the artery for the flow of materials during the construction process and, if they fail, the consequences to the site programme can be significant. They should be considered as early as possible in the process and certainly at the tender stage as there can be considerable financial implications. For example, when building a five-mile stretch of motorway, do you put a temporary road the full length to ensure access throughout its length or do you use existing roadways to access the various structures being built, such as bridges and other parts of the site? On award of the contract the issue should be re-visited, as the information on which decisions were made at the tender stage may have to be amended as more detail becomes available.

Access roads are not just confined to the site, but also include those to be used in bringing materials and people to the site. When constructing the Rakewood Viaduct on the M62, large, heavy steel-plate girders had to be taken along a narrow winding road to access the site. In places this road was not strong enough to cope with these heavy vehicles and had to be strengthened. If any out-sized, long, high, wide or heavy component has to be brought to the site, a route has to be determined. The roads around the site may have restrictions in terms of parking and off-loading, which can have both cost and production implications. Depending on the site activity, it may be necessary to provide wheel-washing facilities to prevent vehicles leaving the site and dirtying the public highway. The public, especially those on foot, need to be protected from being splashed or getting their clothes and footwear dirty when walking past the site.

Consideration should be given to the needs of the personnel working on the contract. It is useful to establish what public transport is available and, for those travelling by car, the parking facilities provided in the local area and on site. If access by public transport is difficult travelling by car is more likely. If car-parking areas are to be provided as part of the contract, consideration should be given to constructing them earlier rather than later. These can also be used for site accommodation and to clean materials storage if the area is large enough.

Other construction work in the close vicinity can have an effect on access to the site. For example, the football stadium in Coventry was awarded to Laing O'Rourke, but the adjoining road works were given to Edmund Nuttall Ltd. Besides delaying normal road users, diversions put in place could impact on deliveries to the stadium unless consultation between the two companies took place.

The police need to be consulted to discover whether others have made any arrangements during the period of construction that would result in access being difficult or impossible. For example marathon runs, festivals, parades, demonstrations and marches.

1.3.1 Design and location of temporary roads

There are several considerations when planning, designing and locating access roads on site. These include:

- One-way flow is preferred to two-way, because the road can be made narrower if passing places at key unloading points are provided. All vehicles enter at the same place and can be checked to see that the load is as stated before being directed to the correct part of the site.
- It was always argued there should be an exit and entrance to the site so if either became blocked the direction of flow could be reversed. However, having only one gate into and out of the site improves security.
- The entrance to the site needs to be positioned to minimise the interruption to the general flow of traffic on the main highways.
- The route should be as short and direct as possible. Factors affecting this are the likelihood of the road being dug up to permit the positioning of underground services or any overhead obstructions, such as a temporary electrical or telephone supply. The phasing of construction can also influence the decision, for example if certain areas of the site have to be handed over completely before the rest of the contact is finished.
- They should be designed so water drains naturally through the thickness of the material, or permits water to run off. The latter could mean the provision of some form of drainage. In any case excess water must be removed or it can combine with soil deposited by vehicular traffic making the road impassable. It should be noted that with the exception of 'tipper' lorries used for disposing of excavation spoil, the majority of vehicles coming onto site are articulated and have difficulty moving over muddy roads, so temporary roads have to be kept relatively clean.
- The design must be sufficient to support the point loads caused by the trailers of articulated vehicles, if these are left on site without the prime mover (the lorry part). On large contracts, concrete may be mixed on site, in which case the hard standings on which the aggregates and sand are stored should be sufficiently strong to support the high point load caused when the vehicle tips its load.
- If the roads to be constructed for the contract are conveniently positioned to coincide with the construction programme, the road foundation could be used instead. Obvious examples of this are on low-rise residential estates.
- From a safety point of view it may be necessary to provide temporary lighting.

- If tower cranes are being used, the temporary road should run within the lifting radius of the crane (section 1.9) so components can be lifted directly into the building or into storage areas.
- Storage areas need to be located adjacent to the access road so goods can be unloaded and stored safely.
- If the site is on either side of a public road and it is necessary to move plant across, it may be necessary to provide traffic control manually or with automatic traffic lights.

Access to the site is not always by road. On rare occasions there is a rail line connected to the main network running into the site, such as at the naval base Marchport at Portsmouth where the navy loaded much of the task force ships on their way to the Falklands. After the war, it was extended enabling the contractor to bring in certain bulk materials by rail. Containers can also be used to deliver materials over longer distances by rail provided the supplier and the site are within approximately an hour of their respective Freightliner depots. The majority of the precast concrete structure and cladding of Gartnavel Hospital in Glasgow was manufactured on the north side of Manchester and about 40 tons was shipped by this means overnight to arrive on site for work the following morning. This was approximately 8 per cent cheaper than using roads.

In Hong Kong barges brought in the aggregate and sand used for some of the large structures constructed on the waterfront. In one case, because of the restricted site space, the mixer set up was also constructed on a moored barge and the mixed concrete brought ashore on conveyors and distributed with concrete pumps.

1.3.2 Materials of construction

There are various materials that can be considered for the construction of temporary roads. Much depends on the frequency of traffic on the road, the type of soil on the site, the climate and the availability of materials. In extremely hot climates, for example, the natural soil may be perfectly adequate for moving vehicles. There is a problem only when there is torrential rain that may occasionally occur, such as during the monsoons. Providing construction only takes place during the dry months, this may be a perfectly adequate solution. Often when it is raining, the precipitation is so great, that construction ceases in any case.

Concrete, sometimes reinforced, whilst expensive, is ideal where the passage of vehicles is excessive. This would be used where the distance is small, such as access from the public road when the footprint of the building covers most of the site as well as around a mixer set up. Hardcore and quarry bottoms (effectively waste material from the quarry of various sizes and shapes that cannot be used for structural purposes) are commonly used for temporary roads.

In areas where there is significant demolition of brick buildings, broken brick can be used. This is a diminishing market in the UK due to the lack of housing replacement, compared with the slum clearance programme of the 1960s, but also because bricks, especially facing bricks, are being recycled.

Geotextiles are also used to strengthen the road base, providing the edges can be restrained. Notably, the fabric Terram allows the passage of surface water through it but prohibits the vertical movement of the soil beneath. Hardcore is then placed on top to complete the roadway.

In severe soft and weak ground conditions, large slabs of expanded polystyrene have been used as a means of cushioning the load, but this would be an exceptional solution. Timber railway sleepers were used in the 1960s as a result of the reductions made to the rail network after the Beeching Report (1961), but today sleepers are usually made from pre-stressed concrete that are unsuitable for access roads. However, there are still parts of the world where they are readily available. Finally, in case of emergencies the Army uses metal roadways. These are in roll-form housed on the back of the vehicle. The leading edge comes over the cab and is laid by driving the lorry forward over it, thereby unrolling the metal sheeting.

1.4 Provision of services

Usually the construction process requires water, electricity, telephone and sewerage services. Other services such as gas and cable services for television may be required for the completed building. These services can come from existing services running close to or adjacent to the site, be brought in by the statutory authorities, or provided by the contractor. Those requiring excavation during instalation will cause some disruption, but if this is planned for it should be minimal, being early in the contract programme.

Water is required for the temporary offices, general use and for the wet construction processes such as the production of mortar, plaster and concrete. In the case of the latter, if mixed on the site, the quantities involved can be considerable and if the rate of supply to the site is inadequate, it may be necessary to have storage facilities on site to guarantee the volume and speed of flow required. On sites were there is no provision, it will be necessary to bring water in tankers or bowsers and store it on site.

Waste water and sewerage disposal can be accomplished by connecting to the existing mains or with the provision of portable lavatory and washing facilities. On a large site the volume to be disposed of is considerable, so access to the mains is highly desirable and may well determine the positioning of this facility, providing adequate fall can be achieved and the existing system is capable of coping with the extra load.

Electricity is required for the offices, to power the plant, equipment and hand tools. For large fixed machines such as tower cranes, this can require a 400-volt three-phase supply which has to be brought in specially by the statutory authority. Generators will have to be provided where local power sources are not available. It should be noted that because of safety, the supply used on the site should be 110-volts single-phase only. For site buildings and fixed lighting a 230-volt single-phase can be used. See the IEE Wiring Regulations, BS7671.

The sophistication of communication methods is changing rapidly and covers a wide range of options from the provision of telephone landlines, mobile phones and broadband. Precisely what is required depends on the size of the contract, the types of management systems in use, whether or not these are site-based or linked to the head office, and the number of personnel working in the site offices.

1.5 Accommodation for contracting staff, sub-contractors and client's representatives

For accommodation there are two prime requirements to be satisfied: first, the minimum construction regulation requirements; and second, to be able to function efficiently as an organisation in managing the contract. The Construction (Health, Safety and Welfare) Regulations 1996 Regulation 22 Schedule 6 cites the welfare facilities to be provided on a construction site. It includes regulations on the provision of sanitary conveniences, washing facilities, drinking water, accommodation for clothing, facilities for changing and the facilities for rest, which would be used for meals, boiling water and a place to go in the event of inclement weather causing a cessation of work.

The location on the site of the accommodation may be limited by the amount of space available and alternative solutions have to be considered. Typical points are:

- Ideally the operatives' accommodation for eating and changing should be as close to the workplace as possible to minimise the loss of productivity due to travelling between the two. For example, if it takes 5 minutes to get to and from each, at the beginning and end of the day, two tea breaks and lunch, the total loss would be 40 minutes.

- On restricted sites, contractors can look for accommodation close to the contract. Although there will be no view of the site, all the services are connected, the space may be greater than can be provided on site and the costs may be less than providing on-site accommodation.
- Many of the hired units can be stacked two units high to save on-site space.
- Providing permission is granted, accommodation can be constructed over the public footpath; but steps must be taken to ensure the safety of the public, the main issue being the supports holding up the building.
- Due to security issues, it is normal to have a security gate and office at the entrance to the site. If the contractor is employing site operatives directly, this is an ideal place to locate the signing on and off clock.
- The cost of installation is a significant proportion of the overall cost of providing accommodation, so once erected it should not have to be moved unless there is no other alternative. Sometimes the site is so restricted that this has to occur. Site staff could use, for example, an underground car park that would usually have minimal services, leaving an area relatively free from continuing construction work.
- Often there are size and shape limitations to the on-site office areas, but bearing in mind the team may be together for some time, it is worthwhile taking account of the needs of the various functions and the frequency at which communication occurs between each. Clearly the higher the frequency, the closer they should be to each other.
- Accommodation may have to be provided for the client's staff such as resident engineers, clerks of works on large prestigious contracts, entertainment and public relation facilities.

Each function has different needs. Without going into all the functions listed, but to give a flavour:

- Planners require a lot of wall space to post their programmes.
- Site managers spend much of their time meeting others, so they should have an office for themselves with an adjoining conference/meeting room or a combination of the two areas.
- Site engineers require a lot of desk space to spread out several drawings at a time and need to access the site quickly and not bring mud on their boots through the whole of the accommodation complex.
- Site managers sometimes prefer to have a view of the site, but this is not a high priority in carrying out their function and is very much a personal decision.

The size and type of desk, chairs and filing capacity has to be looked at. This may become a determining factor in the amount of space the office user requires.

1.6 Material storage and handling

There is a conflict between current thinking of delivering materials 'just in time' (section 6.6) and the traditional way contractors are paid for materials delivered to site within the period of the monthly valuation (*Finance and Control for Construction,* section 14.4). The trend is towards the former, but it is unlikely the delivery of all materials will achieve this objective and materials will still need to be stored on site either at the place of work or in a designated storage area. Even when 'just in time' delivery is reached there will still be occasions when goods will have to be stored because of inclement weather or machine breakdown, such as the crane.

There are advantages and disadvantages of the traditional approach. If the material is there, material control is easier and it is readily available for use. However, the longer it is on site the more likely it is to be damaged or to deteriorate, there is an increased risk of theft, and it may have to be double handled because it is in the wrong place or in a centralised secure holding point, with further transportation and the probability of more damage, resulting in increased costs. There is also a tendency to lose control of the stock especially because more tends to be taken than is actually needed.

1.6.1 Methods of storage

This depends on the value of material, its vulnerability to damage and weathering, and where and when it is required in the construction process. Ideally, materials should be delivered to the site when needed and placed near the operative who is fashioning or fixing it. There are many considerations concerned with material storage and handling. There is an important relationship between the supplier and the contractor in understanding and agreeing the way materials are handled at the factory and how they are best handled on the site. For example, bricks are packaged in lots of approximately 400, held together using metal or plastic strapping and stacked in such a way that there are two horizontal parallel holes which permit the packs to be lifted and transported using a fork lift truck. Cranes attached to the lorry and tower cranes have similar lifting devices that permit the loads to be off-loaded from the vehicle without the need for breaking down the loads.

A completed building comprises a wide range of materials and components that require specific attention to ensure they are not wasted. To give a flavour, Table 1.1 shows suggested categorisations of materials to demonstrate different handling and storage issues.

Issues concerned with storage and handling include the following, all with the aim to reduce waste:

- When unloading certain materials it is imperative to have the proper equipment to ensure safe and secure lifting, such as using lifting beams so vertical lifting can be assured without structurally damaging the component.
- When transporting materials appropriate transportation methods and plant should be used to ensure no damage occurs.
- Storage areas must be clean and level enough to permit proper storage.
- Appropriately designed support structures, racking and spacers (dunnage) should be provided.
- All goods sensitive to damage from different climatic conditions should be protected by covers or housed in a secure cabin.
- Certain goods, such as cement, have a limited shelf life and need to be stored in such a way as to permit the earliest delivered material to be used first. Other goods left in store for a long time can change colour or start to look dirty, for example the edges of vertically stacked cladding components are exposed to the elements, as the units stacked in front do not fully cover the face. Those at the rear of the stack can become discoloured with pollution. When they were fixed into the building they can look different as a result.

Table 1.1 Materials categories

	Examples
Valuable (small items)	Door and window furniture
Consumables	Nails, tie wire, brushes, nuts and bolts, fixings
Short shelf life	Cement
Medium shelf life	Paint, untreated timber, reinforcement steel
Bulk	Bricks, blocks, aggregates, sand, structural steel
Environmental hazards	Fuel, oils
Easily damaged	Plasterboard, polystyrene
Components	Windows, doors, cladding panels

- Certain materials are delivered in bulk, so adequate and appropriate storage has to be provided. This may mean silos, in the case of cement, or constructing bays to segregate the different aggregates to stop cross pollution.
- Certain materials, like diesel oil, are potential environmental polluters if not stored properly. Steps must be taken to ensure that in the event of a leak, the material can be contained without risk of pollution of the ground and watercourses.
- Consideration should also be given to combating theft and vandalism (section 1.12).

1.6.2 Location of storage areas

Valuable items, small consumables and materials with a short shelf life need to be securely stored so that they can be issued when required. The size and amount of storage provision has to be calculated to take account of the delivery schedule. The location of the storage hut needs to be close to the works to reduce the travelling time of operatives collecting the material and positioned by the access road to facilitate deliveries.

Where tower cranes are in use, all significant materials within the lifting capacity of the crane should be stored within the radius of the crane in clearly designated, clean and properly equipped compounds, made secure if necessary. These do not have to be adjacent to the access road, but the nearer they are, the less time it takes for the crane to off-load from the delivery vehicles. There is a strong case for providing a central secure storage area from which goods are distributed in amounts needed for up to three days' production. This helps to control waste.

1.7 Waste disposal

The subject of waste management is covered in Chapter 5. There will always be waste because of the very nature of the work carried out on a construction site. Previously, waste was either buried on the site or taken away in skips and disposed of in landfill sites. In recent years as a result of increasing environmental awareness, legislation and the costs of tipping waste, attention has been drawn to minimising waste and recycling. To carry out the latter it is generally accepted that the most appropriate place to segregate waste is on the construction site.

The various elements to consider when preparing for waste management on site include the following:

- How much and what type of waste is to be planned for?
- Where does the segregation take place?
- How are materials collected and transported to the segregation point? This can involve the provision of chutes to take materials from the upper storeys of a building.
- How are the materials stored on site before disposal?
- What can be recycled and by whom? Does the company have a list of preferred organisations that will collect such waste?
- Are the waste quantity targets being met?

1.8 Site logistics

The quantities of materials delivered to a construction site are immense and should be planned for. Once the contract programme has been produced, the quantity and frequency of deliveries can be assessed. This, then, has to be equated against the amount of space available for storage, the number of vehicles the site can accommodate on the site at any one time and the restrictions on parking on the public roads adjacent to the site. From this information programmes can be produced to schedule the timing of the deliveries, and suppliers and sub-contractors advised accordingly. In the case of frequent and regular deliveries of structural elements or cladding components it may be necessary to use a holding yard for vehicles to smooth out variances due to traffic conditions such as rush hours. The same might apply for deliveries over long distances.

1.9 The location of fixed plant

Spending time on the correct positioning of fixed plant on a site is well worth doing as the costs of relocating can be prohibitive. The main fixed plant on a site are the concrete mixers, hoists and tower cranes. In all of these cases an electricity supply will be needed.

With the development of ready-mixed concrete it is rare to see a mixer set up anywhere other than on large civil engineering contracts, and only then when the costs of bringing in large quantities of concrete some distance make it more economical to mix on site. The positioning of such set ups depends on the type of contract. For example, a motorway contract will probably wish to have it sited near to the middle of its length providing there are good public access roads, whereas on tall structures where a tower crane is used, it will be more appropriate to have the set up within the radius of the crane's jib. Generally, the mixers need to be positioned as close as possible to the main uses of concrete on the contract to reduce transportation time. However, the closer it is to the site entrance the less likely the delivery wagons will get stuck on the access roads.

Hoists can either transport personnel or goods. In either case the positioning is determined by two criteria: access at the base; and the efficient movement of materials on each floor level. Ideally, on most contracts the latter is resolved, when using only one hoist, by positioning the hoist midway along the floor so that goods have the shortest distances to travel. If using two hoists, they may be at the one-quarter and three-quarters points along the length of the building. However, it may not be possible to achieve this because of other building work such as the construction of a podium area.

Free-standing tower cranes and cranes on rails require substantial foundations. Those tied to the building less so, but these and climbing cranes may require structural amendments to the building to take the extra forces exerted by the crane when lifting. Ideally the radius of the crane needs to be such that it can lift any load that needs to be raised to all parts of the building. One of the simplest ways to establish this is, once the weights of components and items have been determined, is to use a sheet of Perspex scaled to match that of the site plan and drawn with circles of different crane radii. This can be moved over the site plan until an optimum position is found (see Figure 1.1). It also has the advantage that once this position has been established, access roads and storage areas can be sited relative to the crane.

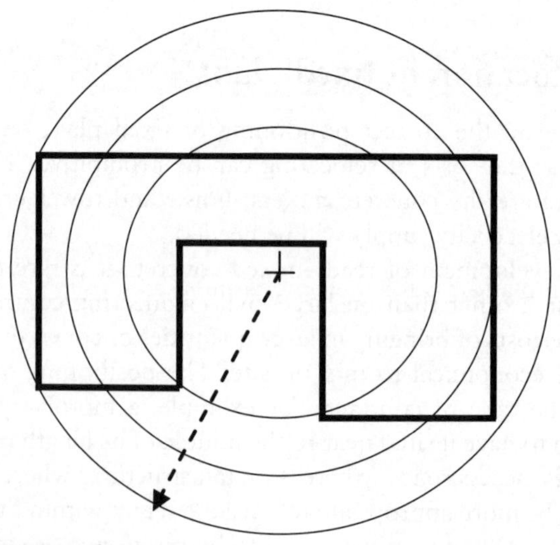

Figure 1.1 Positioning of tower cranes

1.10 Hoardings

Hoardings are the physical barrier between the public and the site activity. Their function primarily is to stop people wandering onto the site and potentially hurting themselves, while also stopping the site activity from spilling over into the public domain. They also provide a first line of defence for security purposes, and if solid, assist in preventing the passage of noise and dust to adjoining properties. They can be used as a marketing tool for the main contractor and, if appropriate, the developer. If there is any danger that objects from the building can fall on the general public, the hoardings have to be designed to give protection. In busy cities this can mean constructing a protected tunnel over the pavement using scaffolding or structural steel. On low-rise housing sites it is often uneconomical to provide hoardings, which means attention to the safety of the public and security become the key issues.

Whatever material is selected to construct the hoardings it should look neat and tidy otherwise it sends out the wrong message to the public and prospective clients. The materials usually chosen are chain-link fencing, plywood attached to timber posts, prefabricated corrugated metal units, prefabricated wire mesh units and the use of existing structures. Chain-link and prefabricated mesh units permits passers by to see into the site so they can report any suspicious activity out of working hours, but also permits a thief to see what is around. On the other hand, plywood and prefabricated corrugated metal sheets do precisely the opposite. The material used may have already been determined by the client in the preliminaries. If plywood is used it will have to be painted either in the company's livery colours or that prescribed by the developer. There have been occasions in the past when the contractor has enlisted the help of the local art college, provided all the paint and allowed each student to create a 'painting' on a plywood sheet. This can generate good publicity and the final results become a local talking point.

The hoardings are normally used to place the company's logo or name at intervals along its length and some contractors provide either windows or observation platforms for the general public to observe the construction process. Some developers require an artist's impression of the final building to be attached to the boundary, which is a good idea as the public is interested in knowing how the building is going to affect their lives and what it is going to look like.

1.11 Communications

Good communications are vital for the effective running of a project and should be considered at a very early stage. Communication on site falls into several different categories (see also *Business Organisation for Construction*, Chapter 11):

- *Advising how to get to the site.* This will include signs on the access roads to the site to direct plant and materials deliveries, or re-routing by diverting site traffic away from sensitive areas, and, on the site, clearly showing the entry points. Maps can be distributed to potential visitors which show parking arrangements, nearest underground and train stations and bus routes. All of these will save them time when accessing or delivering to the site.
- *Communicating with the general public.* Typical examples are apologies for any inconvenience caused, pictures and purpose of the completed building, observation points and the names of organisations involved in the process. These include the developer, architect, surveyors, engineers and main contractor.
- *Communication with the client and its representatives.* Systems, supported by appropriate equipment and software, need to be in place with agreed protocols. This will involve the distribution and control of drawings, contractual matters and points of contact. Similar issues need to be addressed when dealing with sub-contractors and suppliers.
- *Communicating to staff on site and in the head office.* Accomplished with the provision of radio or telephone handsets.
- *Communicating with trades unions.* Where there is union representation on the site, lines of communication should be established and all employees should be aware of the correct procedures.
- *Electronic communication.* More and more information is communicated electronically via email, so having sufficient computer capacity to store and send large amounts of data and an appropriate network system is essential.
- *Health and safety.* Using clear signs to make any official visitor or site employee aware of mandatory requirements, such as hard hats, goggles, footwear, evacuation routes, the whereabouts of dangerous substances, restricted areas, and so on.

1.12 Security

Security is required to reduce, and hopefully prevent, theft, trespass and protest, acts of vandalism and the risk from terrorists. There has always been theft from construction sites and children have found them to be exciting playgrounds. Protests, wanton vandalism and the possible threat of terrorist action have increasingly become issues the industry has to be cognisant of.

1.12.1 Theft

Those thieving can be categorised as professional criminals, on-site staff and, especially on housing sites, the general public. Which plant and materials are the most vulnerable is a complex issue as the proceeds from organised crime are dependent on black-market forces. They can target large moveable plant, such as excavators, or materials in short supply. A few years ago there was a significant world shortage of copper as a result of sanctions against Rhodesia, now Zimbabwe, one of the world's major producers. Building houses without copper pipes was difficult as there was no satisfactory substitute then. A local builder took reasonable precautions, but thieves broke in and stole the fittings. When eventually the builder re-stocked, he parked and demobilised all his plant on all four sides of the hut. The thieves removed the roof.

The theft of wages is a risk and good reason for encouraging operatives to be paid by cheque or directly into their bank accounts. Some of the company's own employees may resort to theft, for their own use or for friends. They may sell information to organised criminals. Having every person searched on leaving the site is expensive, with the alternative being random searches, but in both cases there is the dilemma of how much this would act as a de-motivator, especially if the management staff is exempted. It is important to make it as difficult as possible by having systems in place that actively control the release and return of portable equipment and materials, so as to prevent these occurrences from taking place.

There are two types of theft by members of the public, those who pass by, see something they like and take it, or the residents who live on the estate before building work is completed. It is not unusual for some of these to almost take it as their right to purloin materials. Cases have been cited where residents have built their own garage, garden walls and, in one case, started up the excavator and dumper truck to transport topsoil to their garden. To overcome this problem a contractor posted notices to all the residents offering a reward if a successful prosecution resulted from their reporting an incident. The loss of materials dropped dramatically.

There is a free advice service offered by crime prevention officers, but the quality of this advice depends on the local force and the emphasis it is given. In some cases it could be a police constable, and in others a chief inspector with considerably more experience. Other forces, such as the Greater Manchester Police, have very effective architectural liaison officers who are concerned primarily with designing crime out of buildings and estates, but can also assist in the effectiveness of site layouts. Advice on the positioning of storage areas, lighting and alarm systems is also part of their remit.

Solutions include using lock-ups, alarms, lighting, security patrols outside working hours and security gates and guards during the working day and painting equipment with forensic paint, which is a paint with a composition unique to the purchaser.

1.12.2 Trespassers, vandals, protesters and terrorists

All unauthorised persons who access the site are trespassers. However, the effect that each of the categories can have is different. Those intent on theft have been discussed in section 1.12.1. Others may only use the site to sleep at night, to take drugs, or some may be children exploring and playing games. In all of these cases, the intention is not wittingly to cause damage although they often do. However, if they have an accident whilst on the site, the contractor has a common duty of care and becomes responsible for any injury, unless it can be shown that all reasonable attempts were taken to prevent it. So, for example, removing ladders from the scaffolding would be a reasonable course of action, whereas it would not be if they were left in position where a child could climb them and fall from a height.

Vandalism takes many forms on site, from graffiti, deliberately damaging completed work and unused materials, to arson. Vandals may drive machinery to cause damage or as a joy ride with damage resulting. Besides the costs involved and loss of morale of personnel having to make good, certain acts may make parts of the site unsafe for operatives the following day. In all cases the location of the site will have an effect on the likelihood of it happening.

Some protestors will travel long distances and be prepared for a drawn-out campaign against the development. Normally these sites are predictable either because of potential threat from 'eco-warriors' and the like or because of a build up in the local press reporting on protests about the development long before construction takes place. The problem facing the contractor in the former case, is that they may well be encamped before the contract has been awarded, so the eviction process and the final securing of the site may be a lengthy affair with the need to go through the courts and the need to be sensitive to the opinion of the wider audience, such as occurred at

Manchester Airport's second runway and the Newbury bypass. This makes pre-preparation impossible, unless considered at the development and design stages.

In the past, terrorism has been confined mainly to existing buildings, structures or events, such as the bombing of the Grand Hotel in Brighton (1984) when government ministers were attending the Conservative Party conference. Recently, though, any building being constructed, especially a prestigious one, is vulnerable and could be targeted. Examples of these might have been the Scottish Parliament and the Wembley Football Stadium. Whilst the majority of construction workers are law-abiding citizens, the transient nature of the labour force means that special vigilance needs to be taken, not just in the security of the site, but also in the vetting processes during recruitment.

1.12.3 Security of information

Some of the information about the building may be sensitive. Drawings provided by the Home Office for the construction details of a prison would fall into this category. Protection of all information required for the management and construction of the building also needs protecting against loss and should be considered as part of the business continuity plan (*Business Organisation for Construction*, Chapter 10), as without it the construction work cannot be properly controlled without ensuing delays and extra costs.

1.12.4 Personal security

In certain locations the personal safety of staff, especially female, may be at risk travelling to and from the site. In these circumstances steps should be taken to ensure they are protected against possible verbal or physical attack.

References

Davies, W.H. (1982) *Butterworth Scientific Construction Site Production 4*, Checkbook.

Forster, G. (1995) *Construction Site Studies, Production, Administration and Personnel*, 2nd edn, Longman.

Illingworth, J.R. (1993) *Construction Methods and Planning*, E&FN Spon.

CHAPTER

Contract planning

2.1 Introduction

The act of planning is not confined to industry and commerce. It is part of everyday life, as can be seen every weekend in millions of UK households in the preparation of the Sunday dinner of meat, two vegetables, gravy and a pudding. The meat, vegetables and gravy all take different times to cook and yet all have to be ready at the same time. There is a need to check the progress of the cooking so that times can be adjusted and, on completion of eating the main course, it is expected that the pudding will be ready. The times allowed for each operation will either be found in a recipe or will have been established from experience. Planning the construction of a building follows the same basic principles, but, because of the complexity, requires more sophisticated systems to support the process. Progress checks will also be carried out on several occasions commencing at the development stage, through procurement and the various stages of the construction process.

There is sometimes confusion in the interpretation of the terms planning and programme. Planning is the process of determining the sequence of events or activities that need to occur to complete the project. A programme is the diagrammatic demonstration of the act of planning.

It cannot be stressed enough that a plan must be realistic and therefore attainable, otherwise it is of no use and will fall into disrepute. Used properly it can be a means of engaging the people involved. A managing director of a small civil engineering compan,y whose work comprised mainly laying long lengths of drainage, approached the author who was teaching on a part-time Chartered Institute of Building examination programme. The MD confessed he thought planning was a waste of time and was only taking the course to satisfy the Institute's examination requirements. After a few weeks, he changed his mind and asked if planning could be introduced into his company. After discussion, a weekly programme of works was produced and after the first week, he returned somewhat disillusioned, as only a part of the target had been accomplished. We performed an analysis of the causes for the deficiency of progress and identified some reasons so they could be anticipated in the future. Within a few weeks he was achieving his targets, but not only that, the foreman was now ringing him up and chasing him to ensure deliveries arrived on time otherwise he would not be able to meet the plan.

A good project manager will be able to look at a set of drawings and sketch out an overall programme for the contract, using perhaps only 10 to 20 key activities, but identifying each activity's completion date. The planner will then flesh out the programme and incorporate further sub-activities. Experienced managers argue that the advantage of their programme is that it focuses the rest of the team on key completion dates they have to achieve and that too many activities can distract from this requirement.

2.2 Stages of planning

Planning takes place throughout the process, with the stages and detail depending on the size and complexity of the project:

- the development phase
- the tender stage
- the pre-construction stage
- the construction stage within which there can be various different levels of planning.

Table 2.1, adapted from Griffith *et al.*, (2000), gives an overview of when formal planning is likely to take place. The most common planning techniques used are developed in sections 2.4 and 2.5.

2.2.1 Development planning

When the developer is calculating the financial viability of a potential project (*Finance and Control for Construction*, Chapter 3), a key factor is the duration of the project design and the construction phases, as during these periods no income is being generated and interest has to be paid on the loans taken out to pay for these activities. Further, in retail development, the potential income generation is directly related to the time of year. For example, the run up to Christmas is generally the most profitable part of the year. Who carries out the planning depends on the type of procurement being

Table 2.1 Stages of planning

	Development programme	*Pre-tender programme*	*Master or contract programme*	*Medium (3 months) programme*	*Short-term (1 month) programme*	*Weekly programme*
Small project						
Complex			•	•		•
Not complex			•			•
Medium sized project						
Long duration	•	•	•	•	•	•
Short duration		•	•		•	•
Complex	•	•	•	•	•	•
Not complex		•	•		•	•
Large project						
Long duration	•	•	•	•	•	•
Short duration	•	•	•			•
Complex	•	•	•	•	•	•
Not complex	•	•	•		•	•

adopted for the project. In a traditional form the quantity surveyor is the most likely person, and in management contracting, the project management team would provide this service. The programme would be compiled from limited information, but based on the programmer's experience and the general information available, it is possible to produce a reasonably accurate timescale to complete the project. If this demonstrates that the project as conceived cannot be constructed in the given timescale, modifications to the design can be enacted, the project delayed or abandoned.

2.2.2 Pre-tender planning

This is to assist the tendering process. It is essential the estimator knows whether or not the contract can be built in the prescribed period stated by the client, because if not, then management may consider the project too great a risk and decide not to proceed with the tender. The timing of activities in the programme is required when asking sub-contractors and suppliers to quote, since they need to know what is expected of them before submitting their price. The method on which the programme is based is important as it provides the basis of calculating the activities and preliminaries. It is used at the final tender review meeting to provide information about the cash flow of the project, the amount of risk involved in meeting the programme requirements and any offers that can be made to complete the project earlier than the client has asked.

The prime inputs to the process are from the design team, who produce the drawings, specifications, bills of quantities and contract details, and from the contractor who has the production management expertise and productivity data. This enables the contractor to produce method statements (*Finance and Control for Construction*, Chapter 10) that assist in the production of the programme, site layouts (Chapter 1), the programme and the completed tender documents (*Finance and Control for Construction*, Chapter 10)

The programme will not necessarily be the final outcome as many things can change between this process and the contract being awarded. The design may change and new methods adopted in the light of having more time for consideration.

2.2.3 Master or contract planning

This is sometimes referred to as pre-contract planning as it is produced when the contract has been awarded and prior to the work starting on site. At this point the personnel to run the contract have been selected and may have more developed views on how the works should be executed. This is because they have more information than at the pre-tender stage and can bring different ideas to the table.

They have at their disposal the tender documentation, additional project details from the design team, their own expertise, more accurate data from suppliers and sub-contractors about delivery capabilities and the time they need to complete their part of the work, and the information provided from their site visit. In conjunction with this, schedules of when information is required and key dates for when resources are needed can be established so that orders can be placed with suppliers, sub-contractors and for the package contracts. At the same time delivery schedules of materials and components can be processed. Method statements used for the planning process can be adapted for health and safety purposes (Chapter 4). In doing all of this, the contractor has the opportunity to develop good relationships with all the parties concerned with the construction of the project. Failure to do so could store up trouble for the future because of break-downs in communication.

2.2.4 Contract planning

Throughout the duration of the contract, the planner will continually monitor progress and update the programme as changes in design and delays occur. As indicated in Table 2.1 the length of the contract and its complexity will determine the frequency of producing shorter-term programmes. On long and complex projects, programmes will be produced for three-month periods and updated every two months, and for areas of work that are either complex or critical, programmes of one month and one week will be produced. Whilst there may be valid contractual reasons for the delays permitting extensions to the overall programme, the reality is that many contractors will strive to finish the contract within the original timescale if at all possible to satisfy the client's needs. There may be cost implications in doing this which will have to be resolved with the client.

2.3 Planning and producing a programme

Whatever technique is employed there remain certain fundamentals. The planner has to understand the sequence of operations and their interdependency, compare different methods for accomplishing the tasks in a safe manner, be able to establish the duration of an activity and resource it efficiently. The techniques employed are a function of the complexity of the project. Most important is to remember that the programme is a means of communication to others. Many of the techniques used for planning purposes are almost impossible to interpret by the layman and have to be converted into a readily understandable format. The most usual being a simple bar line (sections 2.4 and 2.5.5).

2.3.1 Calculation of the durations

The planner will establish the duration of the activities from a variety of sources. In the case of the sub-contractor, from their tender document that gives the duration they expect to be on site and amount of notice required before commencement. Alternatively, they will use their own experience based on years of observing similar activities on other contracts or by calculation using standard production outputs and measuring the quantities of materials for each of the defined activities. The latter is not done to the accuracy a quantity surveyor would do when producing bills of quantities, but is accomplished as quickly as possible either by taking the overall dimensions or scaling off from the drawing if not available. Quantities can

Table 2.2 Activity duration calculation

Item	*Quantity*	*Rate*	*Duration*
Overall area, less openings	73m^2		
Labour rate, 2 bricklayers, 1 labourer		0.5 gang hours/m^2 *	
Quantity × rate			36.5 gang hours

* Note the rate includes all the labour costs associated with producing one square metre of wall. It includes for the inner and external skin, the insulation and is inclusive of the laying operation, fixing the insulation and wall ties, fetching and carrying materials, and mixing the mortar.

also be taken from the bills of quantities, but the planner would normally only use the main quantities and ignore the detail. The standard production output data would be sourced from the company's own library of synthetic data (Chapter 3).

Table 2.2 shows a typical calculation for the brickwork from ground floor to first floor level of a detached house assuming an external skin of facing brick, an internal skin of block work with 50mm thick cavity insulation.

In other words, in round figures, it would take one gang one week to complete the first lift of brickwork to this house. Using two gangs could reduce this to half a week, but there is a limit to the number of gangs that could be used in such a limited work area as they could get in each other's way and reduce productivity. One of the planner's skills is to understand this. How many gangs are selected to do the work is determined by the necessity or otherwise to complete the activity faster. For example, there is little point in excavating every foundation on a housing estate within a couple of weeks when the overall contract period is one year as many of the excavations would fill with water and the sides collapse as they weathered, requiring remedial action to be taken.

2.4 Bar charts and linked bar charts

With the exception of line of balance, the majority of programmes will be shown in the format of a bar chart even though many of them will have been underpinned by other planning techniques. They are sometimes referred to as Gantt charts after Henry L. Gantt (1861–1919) who developed them in the 1910s for major infrastructure projects in the USA, such as the Hoover Dam and interstate highways.

These types of presentations used to be produced by the planners using their experience and knowledge. They would calculate how long it would take to carry out each activity, work out the labour required, decide on the sequence of events, consider the time delay necessary before an overlapping activity could start, and then sketch out the programme. They would then establish the amount of labour required in each week and readjust the programme to ensure there was a relatively constant use of the different types of labour. In those days over half the labour was directly employed by the main contractor. They would take the sub-contract activity duration from the sub-contractor and insert it. This method gave an overall idea of how the contract would run, but was fraught with difficulty because of the lack of serious logical thought underpinning the operation However for small contracts where the data are well known, they can work very effectively and be used to monitor progress with some certainty.

Figure 2.1 shows a simple bar chart. The activities are listed sequentially, the first activity being at the top. The bar lines are drawn to scale, the duration having been ascertained as discussed in section 2.3.1. Where activities cannot start before another activity is completed they can be linked together as shown. Where this linkage is used the chart is referred to as a linked bar chart.

Shading in the bar lines each week proportional to the amount of work done is used to monitor progress. By drawing a vertical line, shown in the figure by a broken line, at the current date demonstrates whether or not an activity is behind, on schedule or ahead of programme. In practice simply moving a length of string along the date line can do this.

A traditionally constructed bar line is limited by the fact that it is not easy to reschedule if something needs to be changed as it takes many man-hours to carry it out. The number of activities that can be used is also limited. It will normally from 30 to 60, although sometimes they are produced with as many as 100.

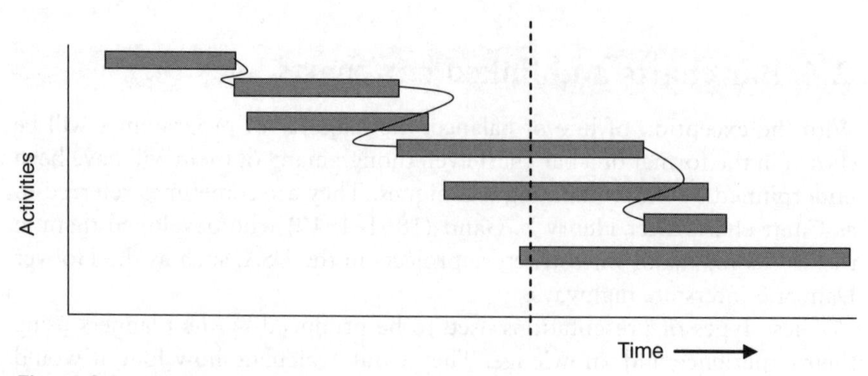

Figure 2.1 Linked bar chart

2.5 Networks

Networks were first developed in the United States in the 1950s when the U.S. Navy Special Projects Office and its consultants devised a new planning scheme for special weapons systems. The outcome was the Program Evaluation Review Technique now known as PERT. Since the duration of many of the activities needed in designing new systems were unknown this system was particularly concerned with assessing the probability of how long they might take. In construction, activities are more readily defined so their duration is predictable, so PERT was found not to be appropriate in the industry. However, a new activity-oriented system was developed for the industry called the Critical Path Method (CPM) or Critical Path Analysis (CPA).

These were first introduced into the UK construction industry during the 1960s to take advantage of the computer technology being bought by the large construction companies. Unfortunately, these computers were relatively slow, the input was laborious and often the output was a network diagram that stretched around the walls of the planner's office and was incomprehensible to all but a few and of little use as a control document to those on site. Now that computers are powerful and fast, large quantities of data can be stored, manipulated and modified at an instant to produce many different outputs, such as the programme and resource implications.

The concept behind the production of networks is establishing the logic of a sequence of events. With bar charts there is a tendency to start with the first activity and decide which activities follow. However, this is not the case with networks. It is a golden rule to remember that 'Which activity comes next?' is irrelevant in producing networks. To ensure the logic is correct the question must always be: 'What activity or activities have to be completed before this activity can commence?'

There are two main ways of producing a network. These are called arrow diagrams and precedence diagrams. Each has advantages and disadvantages, but produce the same outcome so the choice is personal.

2.5.1 Arrow diagrams

An activity can be a combination of smaller activities such as 'reinforced concrete frame', which includes several operations such as form kickers, fix reinforcement, fix and strike formwork and pour concrete. What is included depends on the detail necessary. An arrow is used to represent an activity or operation each of which must have a clearly defined start and finish, as shown in Figure 2.2.

Pour concrete A

Figure 2.2 Arrow diagram

The length of the arrow is not related to the duration of the activity. Indeed it is not unusual to find long arrows representing rather short durations and vice versa. At the two ends of the arrow is a point in time. These are referred to as nodes or events and are represented by circles, as shown in Figure 2.3.

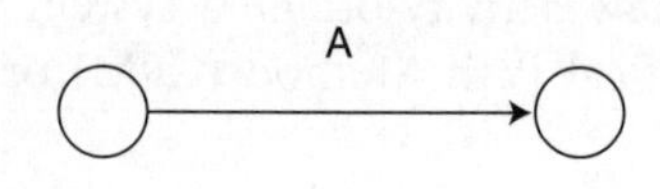

Figure 2.3 Arrows with nodes

Figure 2.4 demonstrates a simple example of the construction of a network. In this case activity B commences when activity A has been completed.

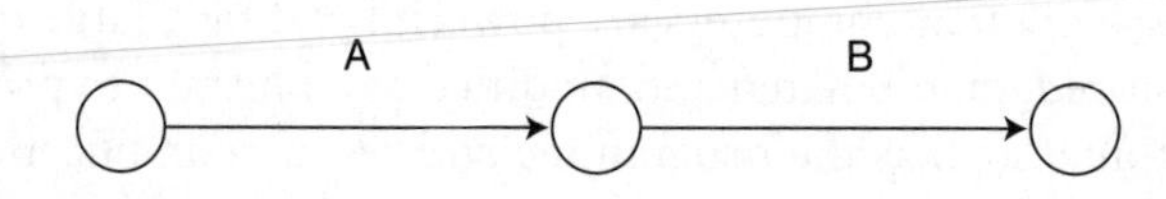

Figure 2.4 Connecting activities

Not all activities are dependent on the completion of just one other. In Figure 2.5 activities B and C can only commence when both activities A and D have been completed.

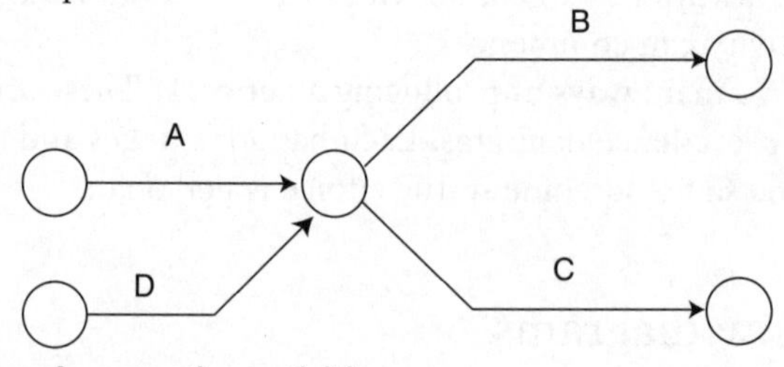

Figure 2.5 Several connecting activities

Note also that activities B, C and D are drawn with part of the arrow horizontal and the description, in this case B, C and D, written adjacent to it. Whilst this is not mandatory, it is recommended as it eliminates confusion with 'dummy' activities (Figure 2.6), which would normally be drawn at an angle.

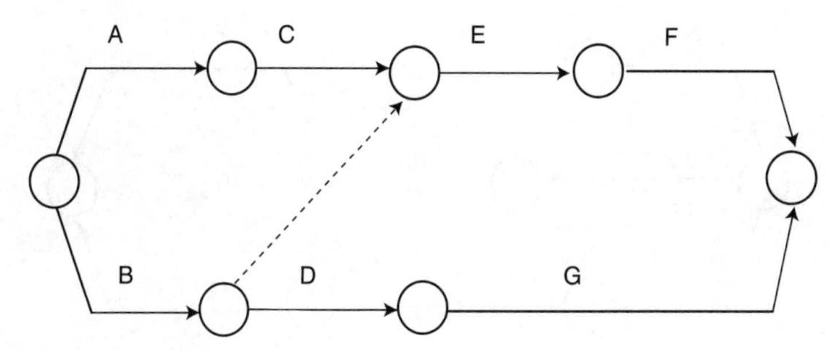

Figure 2.6 Network with dummy activity

Table 2.3 Activity precedence

Activity	*Preceding activity(s)*
A	Start
B	Start
C	A
D	B
E	B and C
F	E
G	D

With arrow diagrams a problem occurs when, as shown in Figure 2.6, activity E can only commence when activities B and C are completed. To overcome this problem the 'dummy' activity is introduced. In this case it is shown between the end of activities B and C drawn at an angle and with a broken line so as to distinguish it from an activity arrow. A dummy has no duration and is used purely as a mechanism to permit the logic to be shown. The information from which the network is drawn is shown in Table 2.3.

In Figure 2.7, activity F can only commence when activities B and D are completed, and activity G when B and E are completed. The information to draw this network is shown in Table 2.4.

An analogy that sometimes assists those learning how to draw arrow diagrams is to imagine that the network represents water pipes, with the water flowing from the start to the finish, the nodes acting as valves. When a valve is opened water can flow, so in the example above, water enters pipe F from both D and B, and enters G from B and E. Therefore the network is drawn correctly.

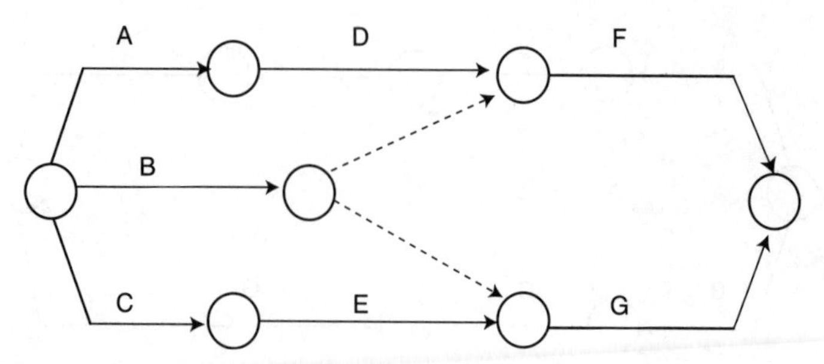

Figure 2.7 Network with two dummy activities

Table 2.4 Activities for Figure 2.7

Activity	*Preceding activity(s)*
A	Start
B	Start
C	Start
D	A
E	C
F	B and D
G	B and E

2.5.2 Definitions

Duration

This is the time that an activity is calculated to take to complete (section 2.3.1). It may be necessary to change the time of this duration when the network is being resourced (section 2.7), normally by reducing the time it takes by increasing the labour content, working overtime or changing the method of work.

Earliest starting time (EST)

This is the earliest time an activity can start within the network. The first activity in network is at the start, so the EST is zero unless the network is being produced for a later part of the contract when it would be from that point in time. For the following activities the earliest the activity can start is the earliest the longest previous activity that it is dependent on can start, plus each duration. So, in Figure 2.8 if the EST for activity A is 0 and its

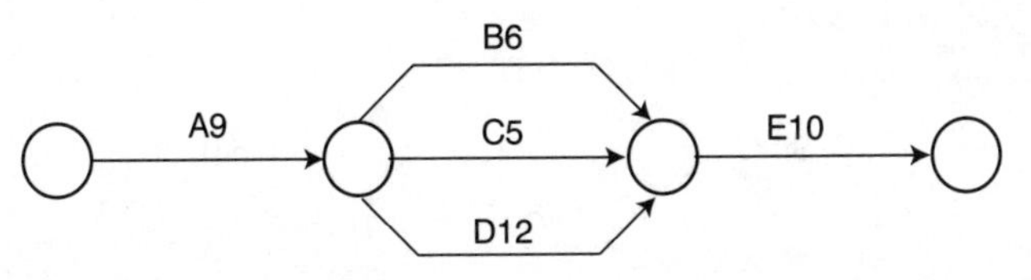

Figure 2.8 Network with durations

duration is 9 days then if the EST for the following activities B, C and D is 9. However, if the activity is preceded by more than one activity the process is more complicated. Activity E can only commence when B, C and D, all of which start at the completion of A, have been completed. Clearly E can only commence when the activity of the longest duration is completed, i.e. D

So, what about B and C which are of shorter duration? Since D takes 12 days, B has 6 days spare and C has 7. This is referred to as float and gives the planner various options. These are:

- Either or both B and C can start at the completion of A and if they begin to go behind programme it does not matter unless this delay is more than 6 and 7 days respectively.
- The start time can be delayed up to and until days 6 and 7 respectively and provided they are completed in 6 and 5 days the programme will be maintained.
- B can be started and then followed by C or vice-versa. This is a decision the planner might take if, for example, both activities required the same piece of plant, as this would be more resource efficient than having two identical pieces on site at the same time.

Earliest finishing time (EFT)

This is the earliest time that an activity can finish. It is calculated as EST plus its duration. So, using the example in Figure 2.8 the EFT for A is 9, the EFTs for B, C and D are 15, 14 and 21 respectively and, because E can only start when D has been completed, its EFT is 31 (i.e. 21+10).

Latest finishing time (LFT)

This is the latest time an activity can finish without affecting the final completion time of the project. In Figure 2.9, it is assumed there are only two activities both starting at the same time, one with duration of 6 days and the other 8 days. The latest finishing time both can end without delaying the contract or further activities is 8.

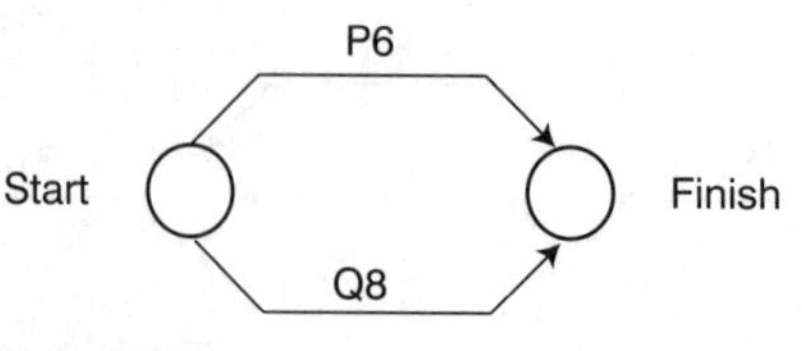

Figure 2.9 Latest finishing times

Latest starting time (LST)

This is the latest time an activity can commence without affecting the final completion time of the project. This is the latest finishing time minus the duration of the activity. In Figure 2.9, activity P could commence 2 days after the commencement of the contract and still not cause delays, but Q must commence at the start.

Some planners and other texts refer to the terms EST and LFT as the Earliest Event Time and Latest Event Time; this is because the point in time represented by the circle in the drawn network is called either a node or an event.

The node or event

As indicated previously, the node is a point in time between the heads of all activities which must finish there in the logic, and the tails of the continuing or following activities which commence from there. This point is used in the calculation process when establishing the critical path through the network. The event number is given to this point in time so that each activity and the dummies can be given a unique reference. This becomes more relevant when using computer network programmes. There are various ways of dividing the circle. For the purposes of this text the format shown in Figure 2.10 will be used, but some writers divide the node into four sections using diagonal lines and insert the EST, EFT, LST and LFT in the quadrants.

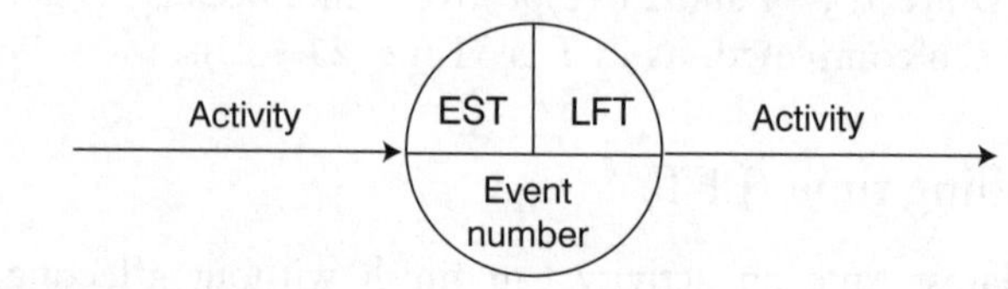

Figure 2.10 The node or event

The critical path

This is the pathway, or pathways, through the network that have no excess time associated with them. In other words, if any of these 'critical' activities are not completed within the duration they have been given, the overall programme will not be achieved and delays will occur unless critical activities further along the pathway are shortened by the same amount. In this case, other activities not previously critical may become so.

Float

On analysis of the network many of the activities will not be on the critical path. The time difference between the EFT and the LFT is the amount of float available. This means that the activity can commence after the EST by this amount providing subsequent dependent activities are delayed by the same amount. This permits the planner to resource the network. Float can be defined in different ways, as shown in Figure 2.11. It demonstrates the relationships of float between two dependent activities A followed by B. The EST, LST, EFT and LFT are shown for activity A, and the EST for activity B.

Total float is the total amount of float that the activity can be delayed without holding up the subsequent activity in the network logic. It is the amount of float that can be used up before it becomes a critical activity. It is therefore the difference between the EFT and the LFT.

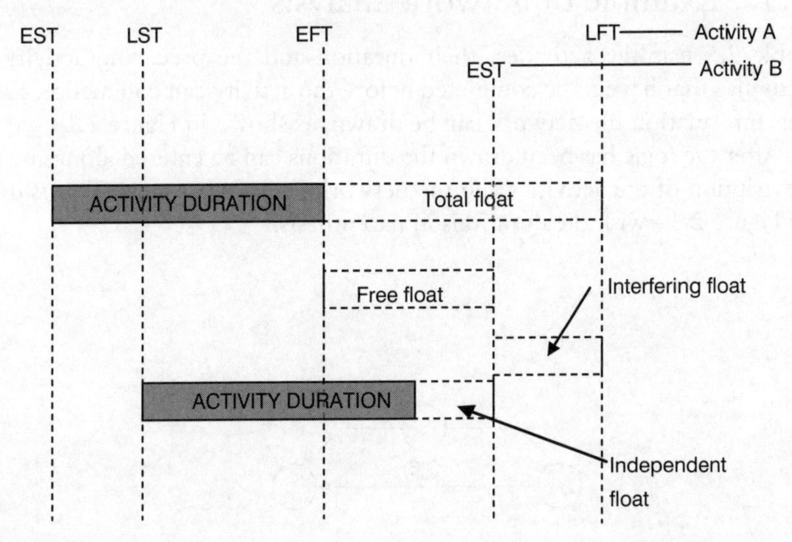

Figure 2.11 Types of float

Free float is the maximum amount of float that is available if activity A finishes at the EFT and the next activity, B, starts at its EST.

Interfering float is the difference between total and free float and is rarely used in any calculation when analysing the network.

Independent float is the float available if activity A commences at its LST and activity B starts at its EST. This is also seldom used in calculations.

2.5.3 Network analysis

The purpose of the network analysis is to determine the total duration of the project based on the durations of the activities, to identify which are critical, and establish the amount of float on the non-critical ones. If on completion the total time is in excess of that required, it can be further analysed and critical activities modified to comply. Those selected will be chosen based on practicality, reality and costs. For example, an activity may be too short a duration to be considered or too expensive to accelerate. There is a limit to the amount of time a critical activity can be reduced as by doing so it may make another critical. For example, as shown in Figure 2.12, where activities A, B and C are 2, 3 and 6 days respectively, it can be seen that it takes 5 days to complete activities A and B, one day less than the critical activity C. If C is reduced to 5 days, then activities A and B also become critical.

2.5.4 Example of network analysis

Table 2.5 lists the activities, their duration and the preceding activity or activities that have to be completed before the activity can commence. From this information the network can be drawn, as shown in Figure 2.13.

After the logic has been drawn the durations can be entered alongside the description of the activities and the network analysis can start. This is done in Figure 2.14 with the durations in parenthesis.

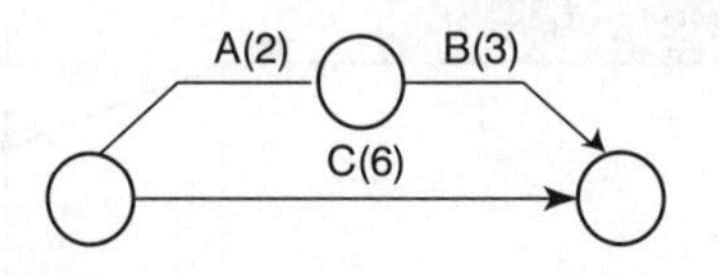

Figure 2.12 Critical path

Table 2.5 Activity data for network drawn in Figure 2.13

Activity	*Duration*	*Preceding activity*
A	10	Start
B	11	Start
C	8	Start
D	6	B
E	3	A,B
F	7	C
G	2	D
H	6	E
J	3	F, M
K	5	E,G
L	4	J
M	2	D

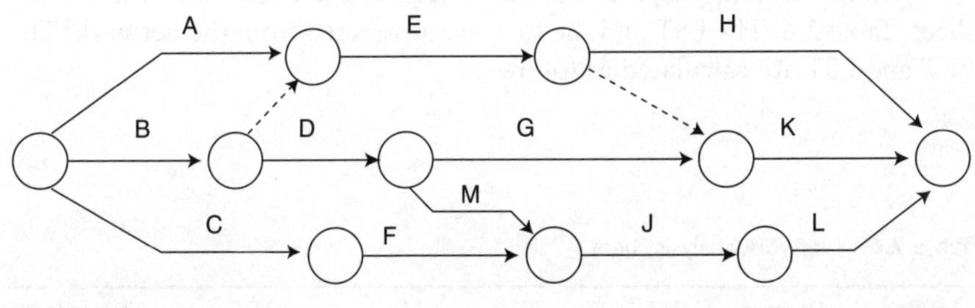

Figure 2.13 Network

To carry out the calculation, first the upper left quadrant of each node is filled in. These are the earliest start times. They are calculated starting on the left side, which is the start of the project, and finish at the right, which is the end. The start is zero which means that the earliest that D can start is when B is completed, i.e. after 11 days, similarly F after 8 days. In the case of E, it can only start when both A and B are completed. Since B takes longer than A, E can only commence after 11 days. Since the earliest D can commence is at day 11 and it takes 6 days to complete, then the earliest activities both G and M can start is 17 days. This process is continued until the finish point where it can be seen that the earliest the completion of the project can be reached is after 26 days.

The second part of the calculation is to complete the upper right quadrant. In this case the pass is made from right to left commencing at the finish. If

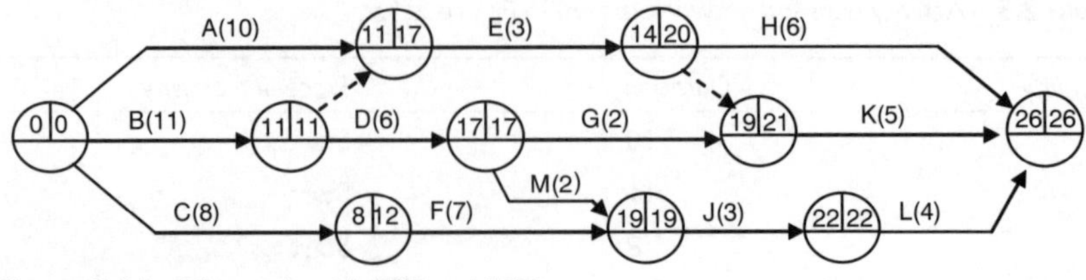

Figure 2.14 Network with EST and LFT

the earliest the finish can be attained is 26 days, then the latest the last activities (H, K and L) must finish is also 26, so this entered in the finish node. Thus the latest time J can finish and still permit L to finish by day 26 is 4 days earlier, i.e. day 22. In the case of E, it has to be finished before both H and K can commence, this means that the latest it can finish is to permit the longest activity to start, i.e. H which takes 6 days is 20 days. This process continues back to the start. It is important to note that the latest finishing time of the start is 0 and if this is not achieved in the calculation there is a mistake somewhere.

When this is complete, the data can be transferred to the network analysis sheet, Table 2.6. The EST and the LFT are abstracted from the network. The EFT and LST are calculated as follows:

Table 2.6 Network analysis sheet

Activity	*Duration*	*EST*	*EFT*	*LST*	*LFT*	*Total Float*
A	10	0	10	7	17	7
B	11	0	11	0	11	0 – critical
C	8	0	8	4	12	4
D	6	11	17	11	17	0 – critical
E	3	11	14	17	20	6
F	7	8	15	12	19	4
G	2	17	19	19	21	2
H	6	14	20	20	26	6
J	3	19	22	19	22	0 – critical
K	5	19	24	21	26	2
L	4	22	26	22	26	0 – critical
M	2	17	19	17	19	0 – critical

EST + Duration = EFT
LFT – Duration = LST

On inspection of the analysis sheet the total float can be calculated in two ways:

Total float = LST – EST and
Total float = LFT – EFT

If these equations don't give the same result there is an error somewhere in the calculation of the EST and LFT on the logic network diagram. Where the total float is zero, then these activities are identified as being critical.

2.5.5 Producing a bar line from a network diagram

A bar line is a relatively simple way of communicating the sequence of the construction activities to others, whereas a network diagram would be meaningless to most, so it is usual to convert the network accordingly. This is a relatively simple operation after the network analysis, but it is important to link activities together that are dependent on one another. From the analysis a line can be plotted showing its EST, EFT, LST and LFT as shown in Figure 2.15. The shaded area represents the duration of the activity and the dash lined box, the total float available.

This means that the activity can start at any point between the EST and the LST without affecting the completion of the contract on time, unlike a critical activity where there is no float. The proviso to this is that another activity is not dependent upon it. For example, as shown in Figure 2.16, assuming that C is critical and both A and B each have float. A can only be started later if B is also delayed by the same amount, because it is dependent on A being completed before it can start.

Using the data shown in the network analysis sheet (Table 2.6) a bar line can be constructed as shown in Figure 2.17.

Activities B, D, M, L and J are those lying on the critical path. They have no float as seen on the diagram and are all dependent on one another in

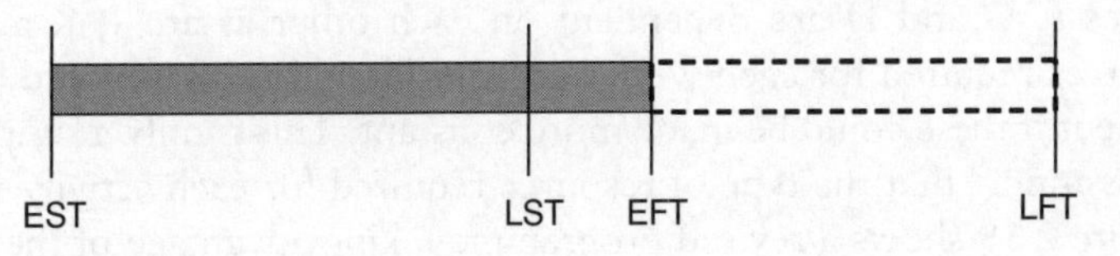

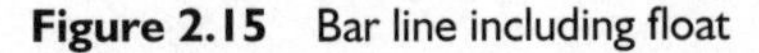
Figure 2.15 Bar line including float

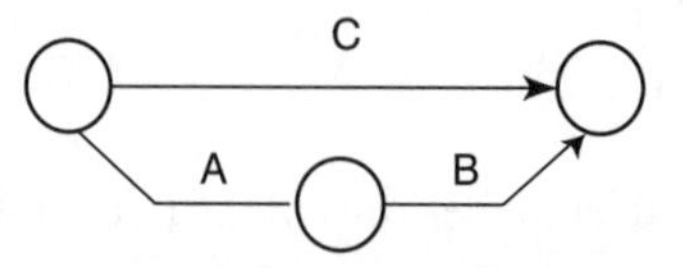

Figure 2.16 Network: C is critical and both A and B each have float

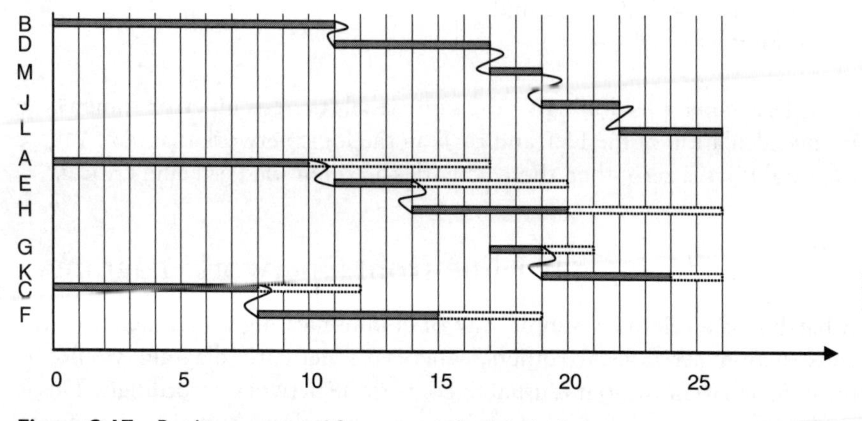

Figure 2.17 Bar line converted from network analysis data

the sequence. All remaining activities have float, but some, such as A and E and E and H, have dependent relationships. Activity A can be commenced one week after the start, but after that cannot be delayed unless E is also. Similarly E cannot be delayed unless H is also.

2.6 Resourcing networks

Once the analysis has taken place on either arrow diagrams or precedence diagrams they can be resourced. Usually, the main considerations are with reducing the overall project time scale, carrying it out with less plant and labour, and trying to do these at the least cost. On the bar line shown in Figure 2.18, the amount of resource required for each week of each activity is entered above the bar line. These are then totalled at the base of the programme. The critical activities A to E have been drawn at the top of the programme, as their start time cannot be moved.

Activities F, G and H are dependent on each other as are J, K and L. The resources required for each week vary considerably and it would be an improvement if these could be made more constant. This is only a simplistic example assuming that the type of resource required for each activity is the same. Figure 2.19 shows a revised programme taking advantage of the float in delaying the start of some activities.

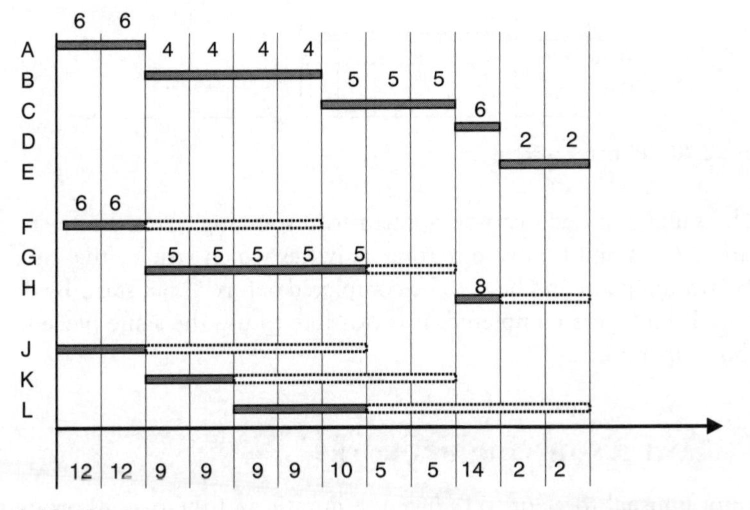

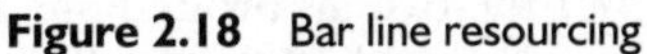
Figure 2.18 Bar line resourcing

By moving some of the activities it is possible to improve the resourcing. It is unlikely that a perfect solution would be achieved in practice. Note that the dependent relationship between activities has been maintained. On a network with many activities it would be time consuming and difficult to achieve the optimum solution. The advantage of programme management computer software is that the various solutions can be produced rapidly for consideration.

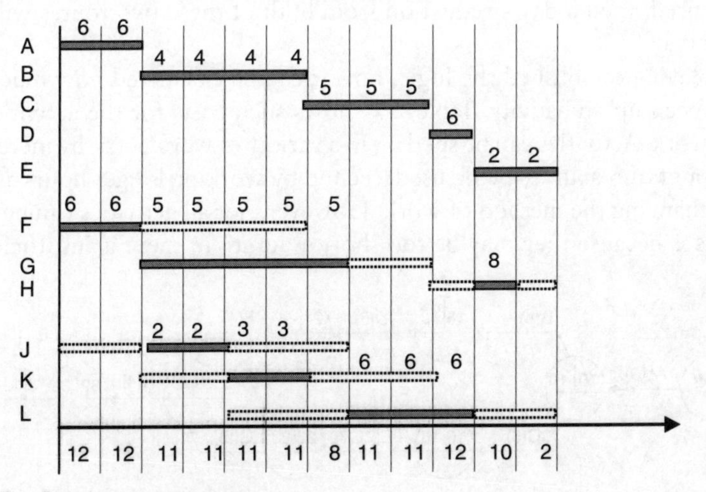

Figure 2.19 Resources programme

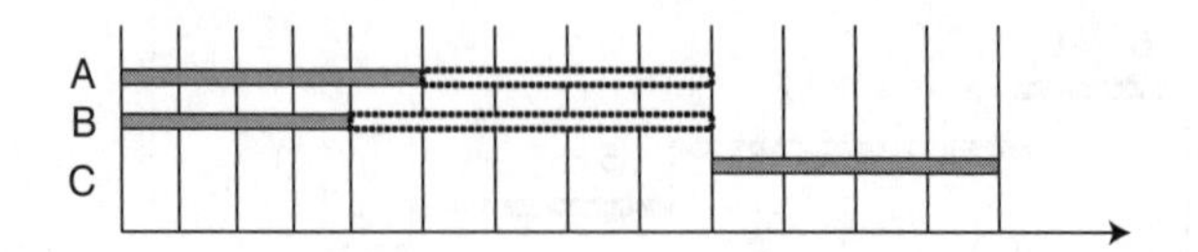

Figure 2.20 Plant resourcing

The same approach can be applied to resourcing plant. For example in Figure 2.20, A and B are excavating activities which require the same piece of excavation plant and have to be completed before C can start. By delaying activity B until A is completed, it is possible to use the same piece of plant for both operations.

2.7 Cost resourcing networks

It is not unusual to want to reduce the duration of the overall programme. This can occur for a variety of reasons: the client wants the building earlier; the contract starts late, but the client still wants the original completion date; and the contractor wants to complete faster than asked for and believes that by offering an earlier completion date when tendering, this may give a competitive edge. Inevitably, an acceleration of the programme will increase the total cost of project. Networks can be used to establish the lowest increase of costs. Figure 2.21 demonstrates a simple network. The critical path is A, C, E, G, H and J and that activities B, D and F have a total of one-day float between them. This means that if any of the activities C, E, or G are reduced by one day then B, D and F also become critical. If another day saving is required, then a day's reduction from both of these two routes will have to be made.

Having established the logic, now it can be calculated how much it costs to speed up an activity. Table 2.7 shows such costs for the activities on the network. Activities can be sped up in a variety of ways. First, by increasing the amount of resources being used, second by working longer hours and finally by changing the method of work. However, not all activities can be reduced. This is because they may be too short a duration, there is insufficient space

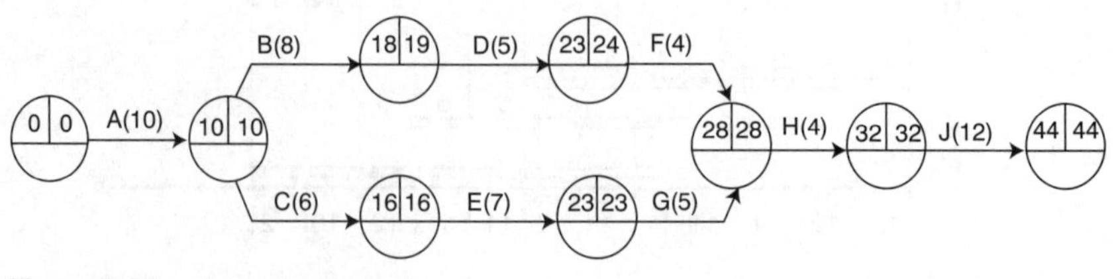

Figure 2.21 A simple network

Table 2.7 Costs of reducing durations

Activity	*Duration*	*Cost of 1-day reduction*	*Cost of 2-day reduction*
A	10	85	400
B	8	63	105
C	6	28	–
D	5	–	–
E	7	102	204
F	4	–	–
G	5	92	–
H	4	–	–
J	12	45	93

available to permit an increase of labour or plant resource, or the length of the activity is already at its minimum. There will also be a limit to the extent that an activity can be reduced and the costs may well become progressively more expensive as the further reductions in time are made.

To establish the cost of reducing the overall programme time the calculation is set out below.

The reduction in cost by one day can be achieved by reducing one of the following activities:

Activity	*Cost*
A	85
B	63
J	45

J is the cheapest. The cost of reducing the overall programme by a further day is from the following activities:

Activity	*Cost*
A	85
B	63
J	93

J has already been reduced by one day hence the cost of the second day reduction is used. In this case the cheapest solution is B. This now means that route B, D and F has now also become a critical path. To save third day:

Activity	Cost
A	85
B + Lowest of C, E, and G which is C	105 + 28 = 133
J	93

Therefore, the cheapest is A. As with the resourcing in section 3.7, to carry out this manually on a large network would be laborious, and project management software packages have made this process much easier.

2.8 Ladder diagrams

The preceding discourse has assumed that as one activity finishes, another dependent one starts, but ignores the fact that in many cases dependent activities commence before the preceding activity has ended. For example, the cladding of a multi-storey reinforced concrete building normally commences long before the structure is completed, as do the various activities involved in laying a long length of drainage. In these cases the activities overlap. This can be overcome by splitting each of the activities into smaller parts. For example, the frame of a 12-storey building could be divided into activities: ground to first, first to fourth and so on. This would work, but makes the network unnecessarily complex and large. An alternative method to show these relationships requires the use of ladder diagrams as shown in Figure 2.22. Here dummies are introduced to show the logic, but unlike before where they were of zero duration (section 2.5.1), in this case they are given a duration. The calculations for the EST and LFT are carried out as before.

The dummy activities 2–4 and 4–6 are referred to as lead time and dummy activities 3–5 and 5–7 as lag time

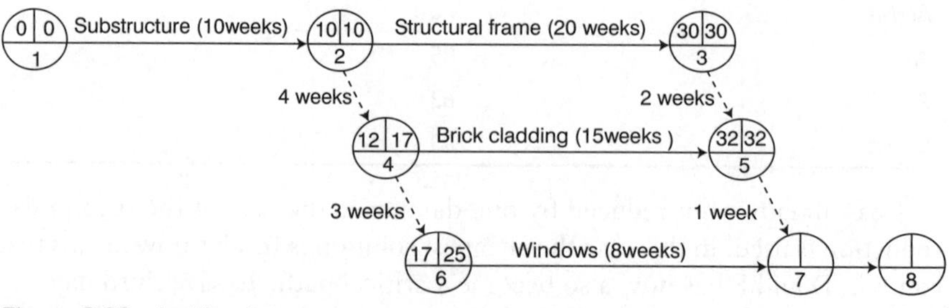

Figure 2.22 Ladder diagram

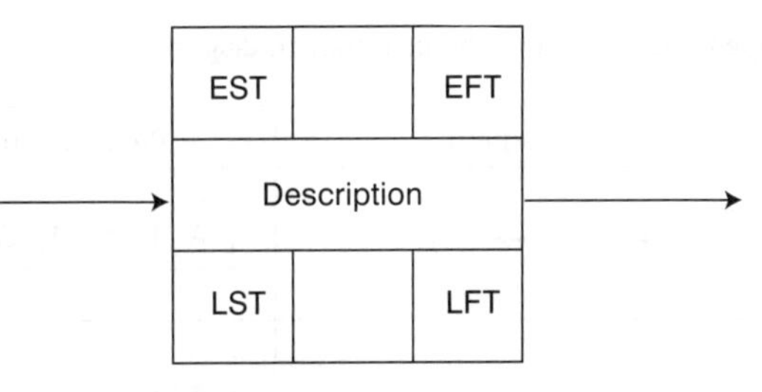

Figure 2.23 Precedence activity box

2.9 Precedence diagrams

The same definitions are used as in arrow diagrams and the same procedure is adopted in producing them. The difference is the activity is shown using a box (Figure 2.23) and the arrow between the activity (box) has zero duration like the node in the arrow diagrams. No dummy activity is required in precedence diagrams and more information is put in the boxes.

Although the logic is identical in both activity and precedence diagrams, their appearance is quite different. Some examples of each are demonstrated in Table 2.8 (overleaf).

An example of a precedence diagram is shown in Figure 2.24. It is constructed from the information shown in Table 2.9. The EST, EFT, LST, LFT, duration and, in this case, total float have been calculated and entered into the activity boxes.

The network analysis is accomplished in exactly the same way as with arrow diagrams and can be manipulated to produce bar line charts and resource finance, plant and labour. However, they can also be used, unlike arrow diagrams, to show relationships other than just one activity must finish before the next one commences. Examples are shown in Table 2.10.

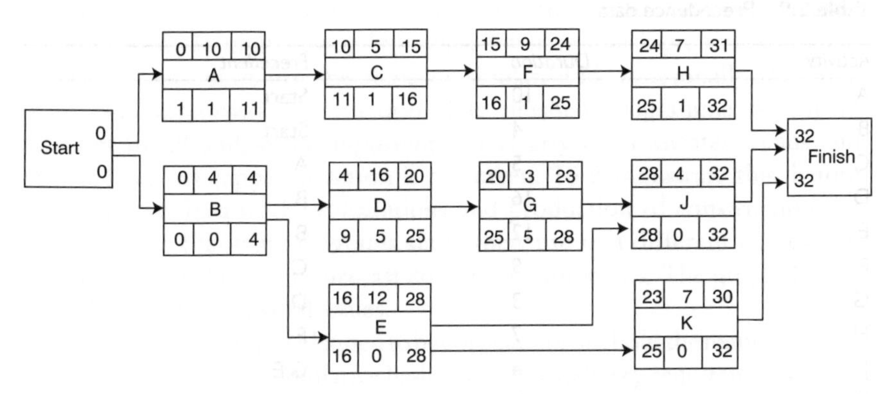

Figure 2.24 Precedence diagram

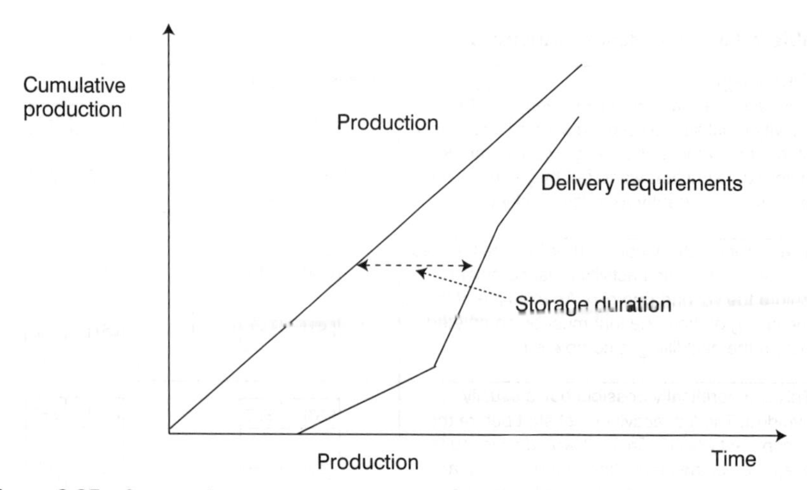

Figure 2.25 Storage duration

the site being supplied increasing its crane capacity, and hence demand, or another contract commencing or finishing. The horizontal distance between the two lines equates to the length of time units are held in stock before being delivered. This is an average as some units will be stored for longer and others shorter depending on the date they were manufactured because the order of manufacturer rarely matches the sequence of delivery. In the case of precast concrete manufacture, this duration is important, as components have to be stored for a minimum time to allow proper curing to take place. What is also clear from this graph is that if production and delivery continue as planned, the factory will eventually run out of units to deliver.

The vertical distance, shown on Figure 2.26, between production and the parallel dotted line equates to the storage capacity at the factory. The cumulative delivery line dips below this for a time. This means that the storage capacity will be exceeded if this were allowed. The solution to this is either to find further storage capacity or to readjust production accordingly, as shown in Figure 2.27.

Figure 2.28 demonstrates how the graph can be used to monitor actual performance. The two dashed lines represent the actual production and delivery requirements. From this information trends can be established. For example, will the loss of production affect the ability to deliver on time? Will the amount of storage space be used up because the delivery requirements are not living up to expectations? And so on.

These examples demonstrate the ability of graphs of this nature to indicate trends to give management the opportunity to take corrective action depending on circumstances. This logic can be applied to the

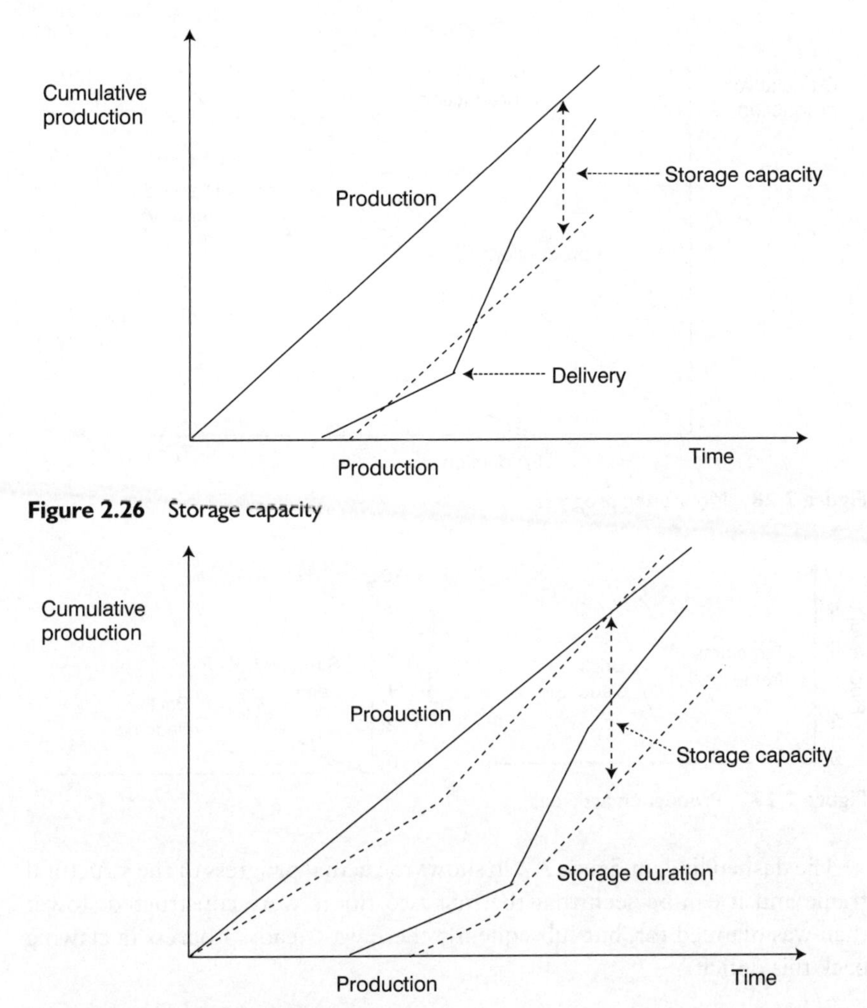

Figure 2.26 Storage capacity

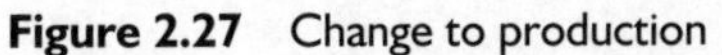

Figure 2.27 Change to production

construction process provided there is sufficient repetition. In this case, shown in Figure 2.29a, the lines shown represent one of the production activities, such as a floor of a reinforced concrete building followed by the brick cladding.

Since the vertical axis reflects the numbers of storeys of the building, it is clear the rate at which the building is to be built. The brickwork starts at the first floor. This is because it is usual to use the ground floor for storage of materials and sometimes as a place to carry out certain pre-construction tasks such as cutting timber to length. When this is no longer required the brickwork from ground to first floor is constructed.

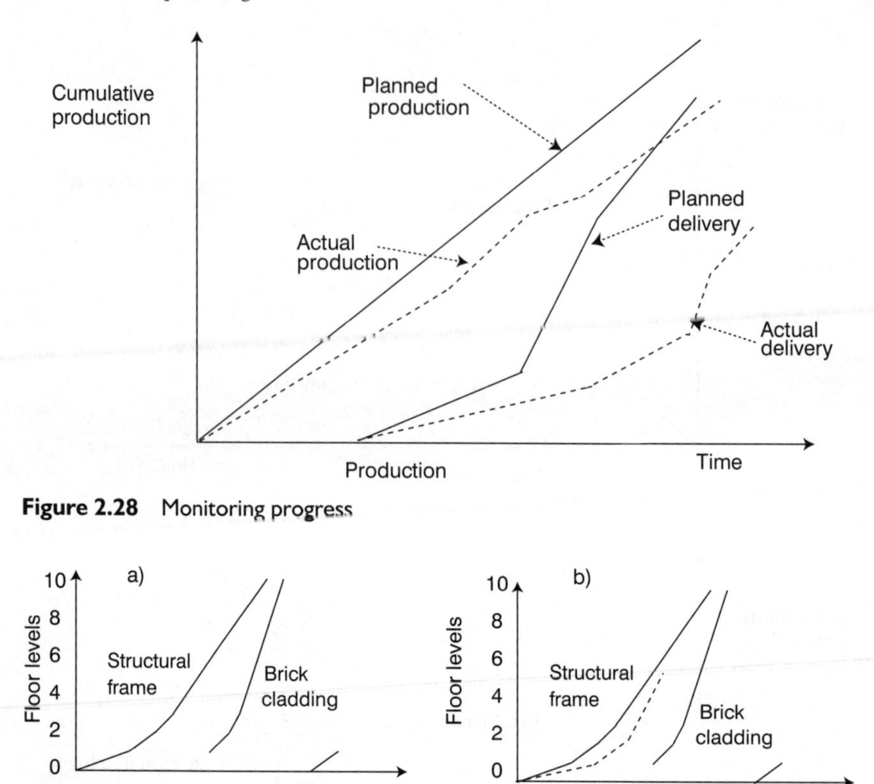

Figure 2.28 Monitoring progress

Figure 2.29 Production activities

The dashed line on figure 2.29b shows the actual progress of the structural frame and it can be seen that the first two floors were constructed slower than was planned for, but subsequently there was steady progress in clawing back this deficit.

2.10.2 Stages in preparing a line of balance diagram

As with any planning, the first task is to identify the activities. In line of balance these activities need to be as few as possible, but cover the main activities. For example, in housing, the simplest would be substructure and ground-floor slab, superstructure, roof structure and covering, finishings and services. This could be broken down into smaller activities, such as ground floor slab, first and second lifts of brickwork, joists, first and second fix plumbing and electrics, tiles and roof structure, trim, and so on.

The next stage is to complete a network analysis of one of the repetitive elements such as in a low-rise residential house and calculate the duration of

Table 2.11 Activities and durations

Activity	*Duration*	*Precedent*	*Days*
Foundations	2	start	
Brickwork	4	foundations	
Roof	1	brickwork	
Window frames, etc.	1	brickwork	
Plumbing	3	brickwork	
Plaster	1	window, roof	
Joinery	1	plaster	
Electrics	1	plaster	
Decoration	2	plumbing, joinery, electrics	
Clean out	1	decoration	

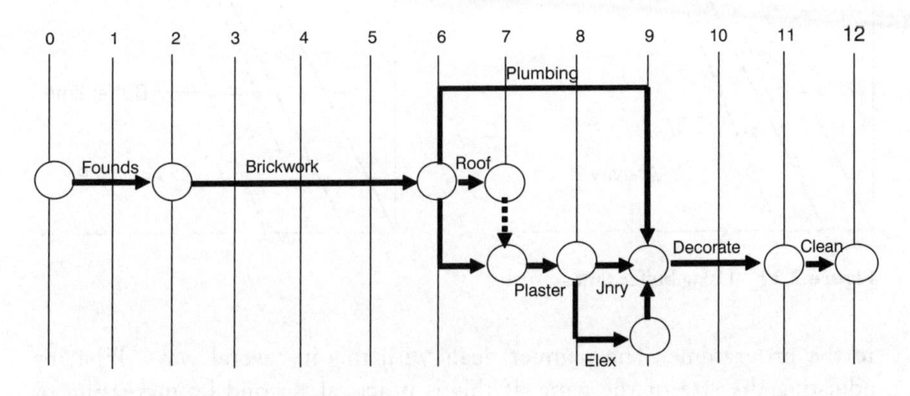

Figure 2.30 Network for typical house

each of the activities. This would take into account gang sizes and the type and amount of plant used. The following worked example, adapted from Pilcher (1992) is based on a typical house comprising the activities shown in Table 2.11 (the number of activates reduced for simplicity) from which the network shown in Figure 2.30 is produced.

For the purposes of this exercise it is assumed there are 200 houses in the estate and the contract duration is 62 weeks. As can be seen from Figure 2.30, it takes 12 weeks to construct one house, so the cumulative graph of the contract is shown in Figure 2.31.

Since each house has a start and finish, the programme can be redrawn as in Figure 2.31b. However, in reality each of the activities needed to construct a house does not take the same amount of time and this has to be dealt with

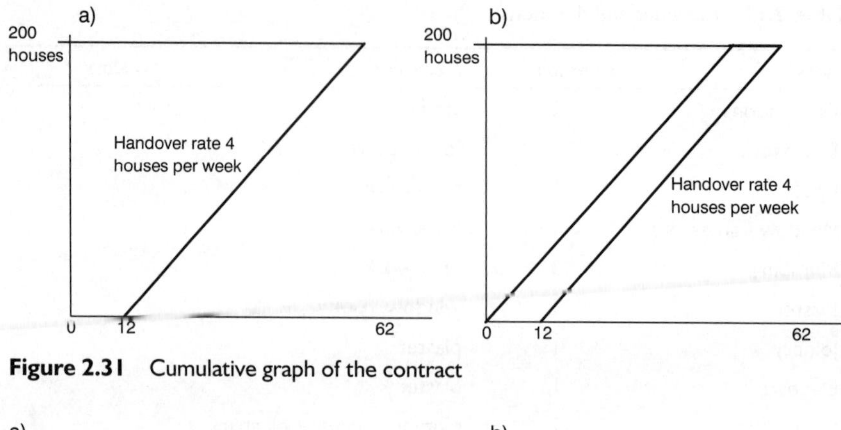

Figure 2.31 Cumulative graph of the contract

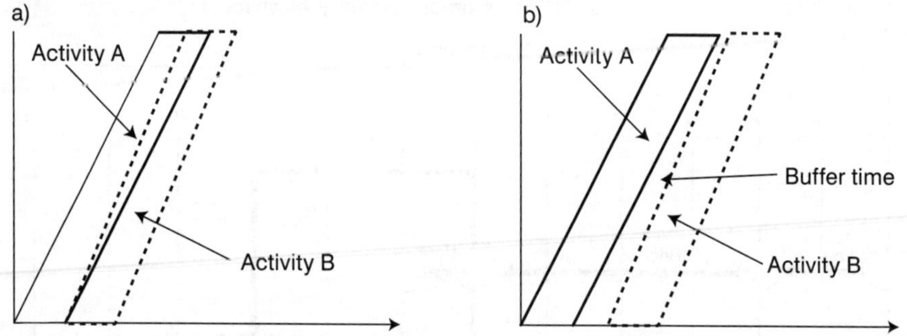

Figure 2.32 Using buffer times

in the programme. The planner deals with this in several ways. First, by adjusting the size of the gang, if this is practical, second by increasing or reducing the number of gangs used, and finally by introducing 'buffer' times to stop interference between activities. Figure 2.32a and 2.32b illustrate before and after utilising a buffer time.

Continuing with the example outlined in Figure 2.30, the outputs for each activity are calculated and resourced so the contract can be completed in the time allocated. The aim is for each of the activities for all the houses to be completed within 50 weeks. This means the slope for all activities needs to be as near to 4 houses per week as possible, and in most cases slightly faster to allow for those that take marginally more, and to accommodate the buffer time. Table 2.12 shows the calculation.

Table 2.12 Line of balance calculation

1	*2*	*3*	*4*	*5*	*6*	*7*	*8*	*9*
Activity	*Estimated man hours/ house*	*Optimum gang size*	*Theoretical gang size*	*Actual gang size*	*Actual output rate*	*Time/day one house*	*Time allowed on slope*	*Buffer time*
	(M)	*(G)*	*(T)*	*(Y)*	*(A)*	*(H)*	*(S)*	*(B)*
Foundations	85	3	9.06	9	3.97	3.78	251	5
Brickwork	200	3	21.33	21	3.94	8.89	252	5
Roof	50	1	5.33	5	3.75	6.67	265	5
Window frames	180	4	19.2	20	4.17	6.00	239	5
Plumbing	140	3	14.93	15	4.02	6.22	248	5
Plaster	40	3	3.2	3	3.75	1.78	265	5
Joinery	120	4	12.8	12	3.75	4.00	265	5
Electrics	90	3	9.6	9	3.75	4.00	265	5
Decoration	110	4	11.73	12	4.09	3.67	243	5
Clean out	20	2	2.13	2	3.48	1.33	286	5

Explanation of the calculations

The target rate of production is 4 houses per week (R)

The normal working week is 37.5 hours

- Column 1 lists the activities as previously defined in the network analysis in Figure 2.30.
- Column 2 is the estimated number of man-hours required to complete each of the activities and is based on production output data.
- Column 3, the optimum gang size, is the usual gang size needed to perform the particular operations effectively. The bricklaying gang is traditionally three, comprising two bricklayers and one labourer.
- Column 4 is the theoretical gang size that is needed to complete the rate of 4 hours per week, i.e. the number of houses required multiplied by the estimated man-hours per hour divided by the hours in the working week.

  $T = (R \times M)/37.5$

- Column 5 is the nearest number of operatives that match both the theoretical gang size (*T*) and the optimum gang size (*G*). For example, in the cases of brickwork and foundations, the theoretical gang size is larger than the actual gang size and for window frames and plumbing is less. This is normal, as it would be unusual if they exactly balanced.
- Column 6, the actual rate of output, is the actual time it would take using the actual gang size rather than the theoretical gang size, i.e. the actual gang size divided by the theoretical gang size multiplied by the target rate, in this case 4.

  $A = (Y \times R)/T$

- Column 7 is the time it takes for each activity for one house based on the actual output rate (*A*). The time taken for this equates to the man-hours per activity divided by the number of men in one gang multiplied by the hours in a working day.

  $H = (M)/(Y \times 7.5)$

- Column 8 is the time in days from the start of the first house to the start of the last house (*S*). This is the number of houses, less one, multiplied by the number of working days in the week divided by the actual rate of build. Figure 2.33 illustrates why one house is deducted from the total. *S* values in the table are rounded up or down to the nearest whole number.

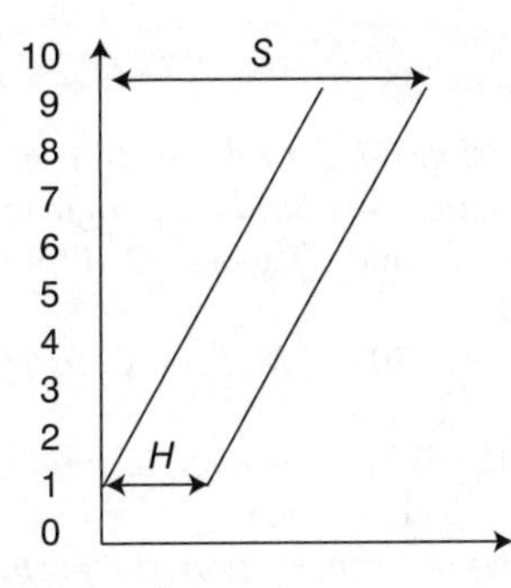

Figure 2.33 Reason for deducting one house

$$S = [(200 - 1) \times 5]/A$$

- Column 9 is the buffer time allowance based on the planner's experience. For simplicity it has been assumed to be the same for each activity, but in practice, this may vary between activities.

From this information the line of balance can be drawn. Table 2.13 demonstrates the calculation to establish the starting and finishing times of the time allowed on the slope line of each activity. The start time of an activity is calculated by adding the buffer time and the activity time for one house (*H*) to the finishing time of all the previous starts and buffer times of activities on the first house. To this is added the time allowed on the slope line (*S*) to give the start time of the last house, in this case rounded to the nearest whole number.

Table 2.13 Starting and finishing times calculation

Activity	*Start Time*	=	*Time allowed on slope line (S)*	*Start time of last house*
Foundations	0	0	251	251
Brickwork	3.87 + 5	8.87	252	261
Roof	8.87 + 8.89 + 5	22.66	265	288
Window frames	22.66 + 6.67 + 5	34.33	239	273*
Etc.				

References

Fisk, E.R. (2000) *Construction Project Administration*, 6th edn, Prentice-Hall.

Forster, G. (1994) *Construction Site Studies*, Longman.

Griffith, A., Stephenson, P. and Watson, P. (2000) *Management Systems for Construction*, Longman.

Harris, F. and McCaffer, R. (2006) *Modern Construction Management*, 6th edn, Blackwell Science.

Oxley, R. and Poskitt, J. (1996) *Management Techniques Applied to the Construction Industry*, 5th edn, Blackwell Science.

Pilcher, R. (1992) *Principles of Construction Management*, McGraw-Hill.

CHAPTER

Work study

3.1 Introduction

To accurately plan work, calculate the labour and plant elements of an estimate, and produce targets for incentive schemes, it is necessary to have data based on the measurement of work. Whilst the study and measurement of work has been considered by many, it was Fredrick Winslow Taylor (*Finance and Control for Construction*, Chapter 1) who is considered the founder of the subject. He broke work activities into discrete elements, timing them with a stopwatch many times to achieve a statistical average. He was the first person to apply a scientific approach to the subject. Frank Gilbreth approached the problem from a different angle by asking the question: 'Is there a better way?' He studied the movements of bricklayers to arrive at what is considered the forerunner of method study.

The concept of 'time and motion' studies were once viewed with suspicion as it was perceived by many as a mechanism for management to make operatives work harder without any increase in pay. This manifests itself when management employs work study engineers to look at a method of work to see if they can improve the time it takes to carry out an operation or to reorganise the work so that it can be carried out with fewer personnel. It is therefore very important that whenever work study is employed, the workforce under scrutiny understands why it is being done, and to eliminate the 'cloak and dagger' image of the process. Initially, it was an activity addressing production issues, but recently this has been extended to encompass maintenance work, clerical and administrative procedures, and other non-manufacturing environments.

The application of work study on construction projects is more difficult due to uncertainties such as the weather, labour turnover and unpredictable problems associated with ground works. Further, as more work has been sub-contracted, much of the onus for carrying out work study falls on these organisations which may not be large enough to sustain the added overhead on what for them is a short-term contract. However, there is still a use for the techniques and, as supply chain management increases in popularity, one of the services the main contractor or contracts manger may do is provide this service. The main contractor is more likely to use method study to improve productivity especially where many trades have to be co-ordinated. The author still finds that Rowland Geary's book, *Work Study Applied to Building*, though published in 1962, is still applicable and some of the examples used in this chapter have been adapted from this text.

The object of work study is to assist management in obtaining the optimum use of human, plant and material resources that are available to an organisation for the accomplishment of the work. It has three aspects:

1. to obtain the most effective use of plant and equipment
2. to employ human effort in the most effective way
3. to evaluate the time and effort required to carry out specific tasks.

It is defined in the British Standard for Work Study BS 3138:1969 as:

> A management service based upon those techniques, particularly method study and work measurement, which are used in the examination of human work in all its contexts, and which lead to the systematic investigation of all the resources and factors which affect the efficiency and economy of the situation being reviewed, in order to effect improvement.

The key issue is that of improvement, which is normally interpreted as higher productivity. There is no point in carrying out what can be an expensive and lengthy process if this is not the intention, although there can be occasional exceptions such as when improving safety or quality.

Work study is divided in to two distinct sections, that of method study and work measurement. However they are inter-related as can be seen in Figure 3.1.

Method study is defined in the 1969 British Standard as:

> The systematic recording and critical examination of the factors and resources involved in existing and proposed ways of doing work, as a means of developing and applying easier and, more effective methods and reducing costs.

and in BS 3138:1992, work measurement as:

> The application of techniques designed to establish the time for a qualified worker to carry out a specified job at a defined level of performance.

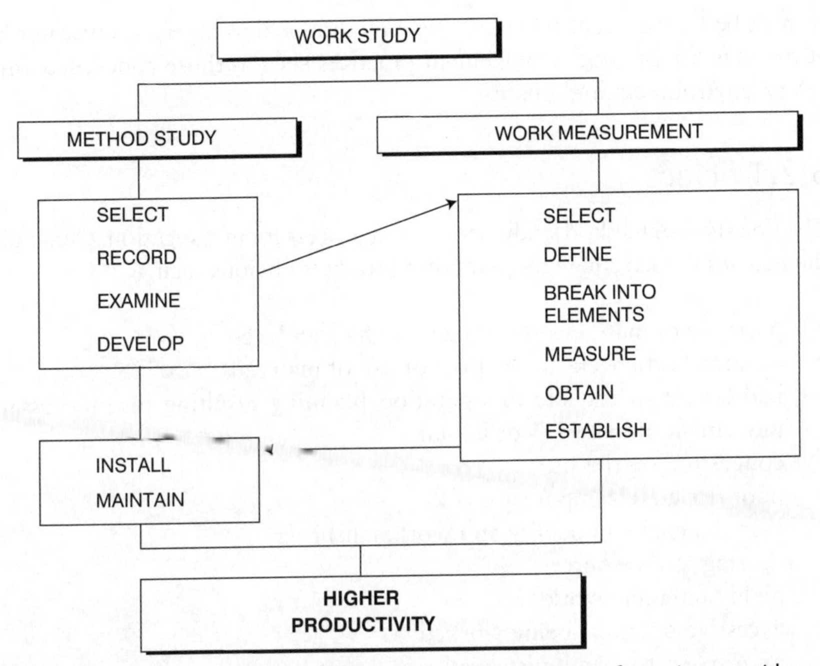

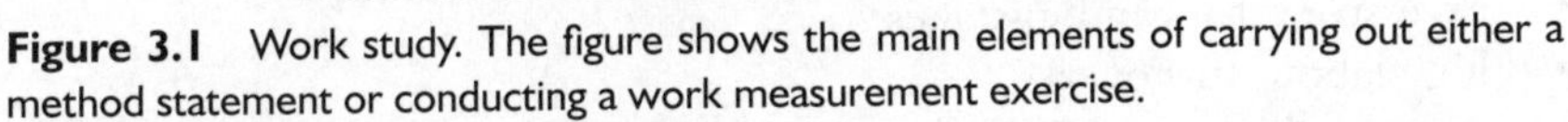

Figure 3.1 Work study. The figure shows the main elements of carrying out either a method statement or conducting a work measurement exercise.

And a qualified worker being is defined as:

> A worker who has acquired the skill, knowledge and other attributes to carry out work in hand to satisfactory standards of safety.

Further, the qualified worker must have been working on the process for a while so that they have gone through the learning curve.

3.2 Method study

All managers involved in the construction processes, whether working for the main contractor or the sub-contractors, should carefully consider the method used in carrying out any tasks. The principles outlined below are employed by the work study engineer but are also the basis on which the manager can operate. A further definition of method statement is:

> Involves the breakdown of an operation into its component elements and their systematic analysis. If on analysis a component cannot withstand investigation then it is either eliminated or improved.

It must be stressed that when any new method is developed, it must not be at the expense of good management practices such as those concerned with safety, environment and quality.

3.2.1 Select

The first stage of method study is to select a process for investigation. Generally, the reasons for selecting an operation to study is obvious such as:

- poor use of materials resulting in high scrap levels
- existing bottlenecks in the flow or use of materials
- bad layout of the site or operation planning resulting in unnecessary movement of materials or labour
- congestion on the site
- poor design of temporary works
- inconsistencies in quality and workmanship
- shortages of resources
- highly fatiguing work
- excessive overtime being worked
- excessive errors and mistakes
- high labour turnover
- poor working conditions
- employee complaints about their work without logical explanations
- delays in obtaining information
- regularly not meeting programme requirements.
- not meeting cost targets
- excessive periods of idle plant
- excessive plant breakdowns.

The two main expectations from carrying out a method study are to increase the levels of production by reorganising the labour and plant to reduce the production costs, and to produce the same or similar output rates, but with less labour or plant. Another potential use is to investigate the method currently employed where there is excessive scrap or waste material with an aim to reduce it. However, less obvious benefits can also be derived, including increasing the quality of the product or service without slowing down the process significantly or using more labour and improving safety conditions in the work place, which can also mean improved standards of cleanliness and housekeeping.

3.2.2 Record

There are several methods of recording the method currently in use. Each has to be selected for the type of work to be recorded, as they are not universally appropriate. For example, in the construction industry most operations take a considerable time to complete, usually by gangs of operatives, rather than individuals. Compare this with the assembly of a computer or a vacuum cleaner where one operative will be putting together several components in a very short time scale. The other main issue in construction is the flow of materials from arrival on site, or storage, to the place of work and through the production process. This may also involve sub-operations not at the work face. For example, pre-assembly of reinforcement steel at ground level or the discharge of ready-mixed concrete into a skip to be transported by crane into the building. The main recording techniques used in construction for recording are charts and flow diagrams.

The process chart is the main chart in use, which diagrammatically demonstrates potential problems in the current work situation. The process is simplified by the use of five symbols, into which all activities can be divided (Table 3.1).

There are two types of flow chart called the outline process chart, which gives an overall picture of the flow using on the symbols for operation and inspection, and the flow process chart giving much more detail, using all five symbols.

An example of a process flow chart taking mixed concrete from a storage hopper, via dumper and crane to the top of the building, where it

Table 3.1 Flow chart symbols

Symbol	*Activity*	*Description*
○	Operation	Where something is produced, accomplished or furthers the process in some way.
⇨	Transport	Where material or persons are transported. The arrow can be pointed in the direction of the transportation activity for clarification.
D	Delay*	This is where a delay, or an interruption or interference with the process occurs.
▽	Storage*	If the material is stored for later use, e.g. a timber store, or it is held, such as one hand holding a bolt whilst the other attaches a washer.
□	Inspection	Where quality or quantity is checked and verified.

* Sometimes the distinction between the symbols for delay and storage are difficult to identify

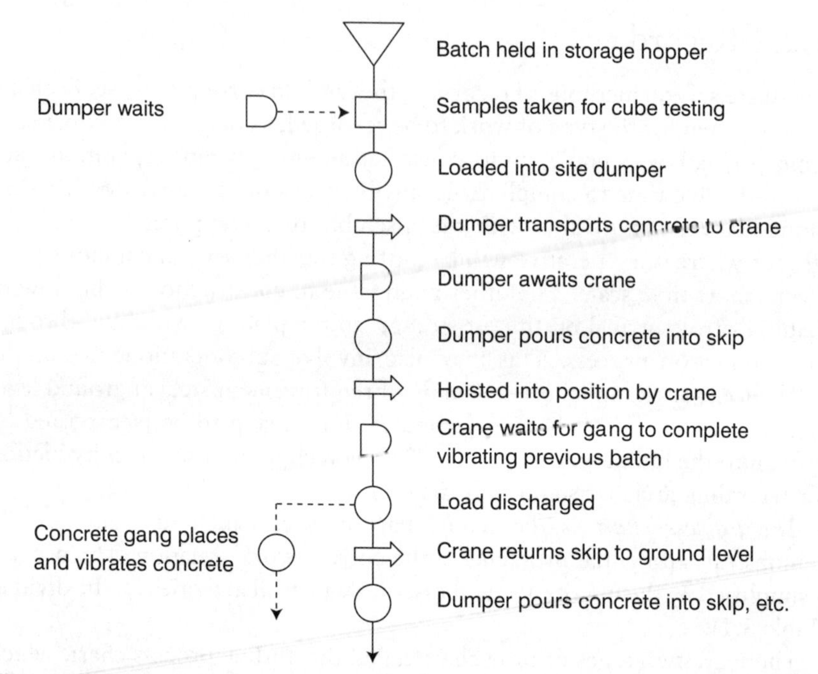

Figure 3.2 Flow diagram

is discharged, placed and vibrated is shown in Figure 3.2. Not shown here, but usually added, are the times each of the 'activities' takes. This helps in the examination process because, for example, if a delay were minimal, then improvement would not be as important than if it was of significant duration.

The information from flow charts can be transposed onto the plan of the process so that the overall flow in relationships to the total workplace can be visualised. Equally, it can be shown on a three-dimensional model showing both the horizontal and vertical movements.

Travel charts can be used to record the movement of people between office locations. Originally intended to record the movement of materials from either one department or machine to another, it is ideal for demonstrating the movement and frequency of personnel from one office to another. In the former, as shown in Figure 3.3, X indicates that an item has moved from 2 to 3. On the other hand using it as a measure of frequency, the 8 indicates there have been 8 travels from room 6 to room 4 and equally that, as would be expected the reverse that there was 8 return trips from room 4 to 6, unless the person went via another room.

Multiple activity charts are ideal for construction work as gangs of operatives carry out most work, and accommodate the recording of multiple

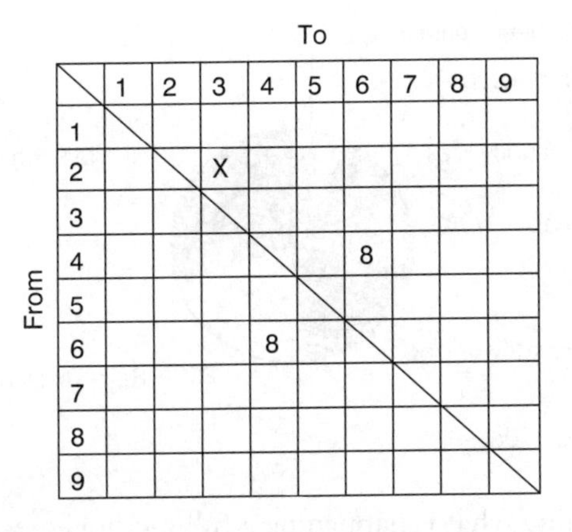

Figure 3.3 Travel chart

tasks. They are also clear to understand. The current operations are timed and normalised using work measurement techniques (section 3.3.3) and are then produced in a bar line format, one axis being time and the other the operatives and any plant or equipment they may be using. The time is usually on the horizontal access, but as demonstrated in section 3.2.5, this is not always the case. Figure 3.4 is a hypothetical example of a multiple activity chart recording a repetitive operation, the shaded areas representing working time and the unshaded, idle time. Once the chart has been produced it can be examined and improved methods developed.

Foreman delay surveys rely on those actually doing the job to have a good idea of what the production problems are. The foreman is asked to record the delays each day, the resources lost as a result, and the causes of the delays. Typical causes of delay include waiting for materials, waiting for tools and equipment, plant breakdown, remedial work whether caused by design or production errors, waiting for information, lack of continuity, moving from one place to another, difficulties of access, and inclement weather.

Time →

	1	2	3	4	5	6	7	8	9	10	11	12	13
Operative 1													
Operative 2													
Operative 3													
Operative 4													
Plant 1													
Plant 2													

Note: Plant 1 is operated by operative 3 and plant 2 by operative 1

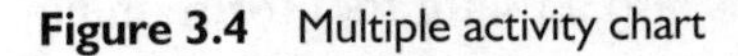
Figure 3.4 Multiple activity chart

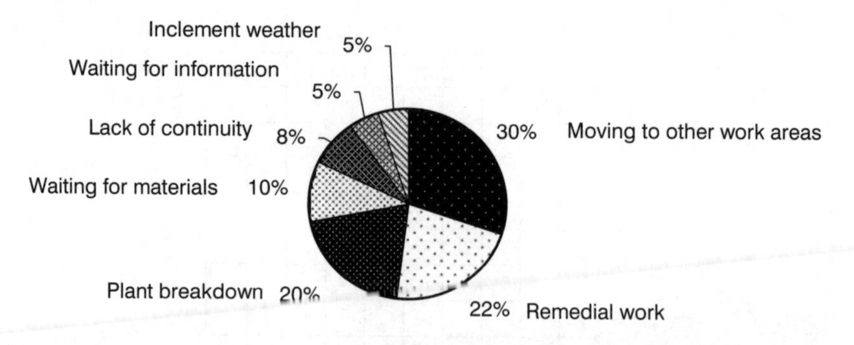

Figure 3.5 Causes of delay

After establishing what is happening a full picture emerges of the relative significance of each delay to the whole as shown in Figure 3.5 so it can be analysed, priories established and action taken after the examination process has taken place.

If each of these percentages are then plotted against the rest of the day's productive work, and regularly checked after improvements have been made, trends to improvement can be monitored. Some of the advantages of this approach are that the foremen are engaged in the process and take interest in eliminating the delays, there is instant feedback of problems, and it is inexpensive to carry out.

There are other methods of recording available, but are less likely to be used for construction processes.

String diagrams are used to record the flow of materials and personnel around the workplace. The engineer records the movement between locations on the site. A scale plan showing the locations is produced and pins placed into the start and destination of the route. A length of string is then wrapped round the pins to show each separate travel. This indicates any excessive movement and backtracking.

Time lapse photography is a useful method of recording activities of medium- and long-term duration which otherwise would take a long time to study. When played back the whole process can be observed in just a few minutes so sequences can be clearly identified. This is an effective method, but in construction, the problem is where to locate the expensive equipment securely.

Slow-motion photography is ideally suited to work carried too fast to record with the naked eye such as the factory assembly of small components. In this case, knowing the rate the film has been slowed down means the work can be examined scientifically. Modern video and digital technology also provide the time scale on the screen. It is also a useful tool for training purposes as it can demonstrate the intricate moves in a manner the trainee can understand and mimic.

Therbligs, which (excluding the 's') is an anagram of Gilbreth, is a system using symbols for different movements of the body, for example the movements of bricklayers. It is unlikely to be used in a site scenario, but could well be used in a research environment.

3.2.3 Examine

The examination process is probably the most important aspect of the whole process as this determines the current situation in detail and from these data, if analysed thoroughly, solutions can be developed. There is a direct correlation between the quality of this process and the final proposals. The reader might bear in mind that this process is appropriate for any problem-solving situation. When examining there are some key issues to remember:

- There is danger that facts are not seen as they are, but as others wish to see them. Therefore they must be examined as they are, not as they might appear to be on the surface.
- It is a failing to approach the analysis with preconceived ideas as to what is wrong and/or what the solution is as this can colour the interpretation of the facts. Any ideas or new methods the engineer might have in advance, should not be ignored, but rather noted down for consideration at the appropriate time.
- It is easy to see the obvious and assume the solution as a result. For example, there may be excessive transport between places of work, but in reality the biggest problem could be delays in the flow of materials at each location caused by inadequate communications between these locations.
- It is reasonable to believe that every part of the process recorded is wrong until proven correct. This means each detail must be examined logically and thoroughly.
- The examination should be approached with an open mind and in a systematic way.

The examination process itself is achieved by using two detailed sets of questions as shown in Table 3.2. These are referred to as the primary and secondary questions, the latter to establish alternatives to the existing or proposed methods.

Each of these questions should be asked in the order demonstrated, otherwise it makes the logical development of ideas difficult, wasteful

Table 3.2 Examination questions

		Primary questions	*Secondary questions*
1	Purpose	*What* is achieved, is it necessary and *why*?	*What* else could have been done?
2	Place	*Where* is it done and *why* there?	*Where* else could it have been done?
3	Sequence	*Who* does it and *why* do they do it?	*When* else could it have been done?
4	Person	*How* is it done and *why* that way?	*Who* else could do it?
5	Means	*How* is it done and *why* that way?	*How* else could it have been done?

or meaningless. Each activity, in the case of the flow diagram, operation, transport, delay, storage and inspection, should be examined using the above questions. In some instances not all of the questions will be relevant. When each of these questions is applied there may be more than one solution. In this case a further examination of the implications must be made before the final choice.

When applying the secondary questions the following considerations should be given. When the purpose of the activity is challenged, the main objective is to see whether or not it can be eliminated entirely. Then if the activity proves essential, the aim is to see whether or not the activity can be modified, either by changing it or combining it with other activities. Finally, the question should be put to see if the activity can be simplified.

Employing brainstorming techniques can assist in the examination process. This technique is explained in more detail in *Finance and Control for Construction*, section 7.3.3, but, in essence, involves a group of people looking at the problem and suggesting alterative solutions without any restrictions of protocol.

3.2.4 Develop and submit

Before developing a new method, the ergonomics of the work situation and the environment in which it is set should be considered, as these can assist in producing a more effective solution. These are:

- *Sight and light* are linked in the sense that the more detailed the work the greater relevance the appropriate level and type of light becomes. If wrong the operative will perform at a lower level of output because of the higher concentration required and can experience eye strain needing relaxation time to recuperate. A clear unobstructed view of the work, tools and materials used is also to be aimed for.

- *Ventilation and heating* levels affect productivity. The temperature required is dependent on the type of work. For example, typists working in an office will require a higher temperature than manual workers because of the different levels of physical energy expended. The number of fresh air changes is equally important as they can affect the general health of staff. At one end of the scale the atmosphere feels heavy and muggy with high relative humidity, and at the other end the worker is subjected to draughts. At the same time any fumes generated need to be dispersed.
- *Colour* can be used to great effect for identification purposes hence speeding up the operation. This can be used for safety reasons such as the colour code used for identifying building service pipes so that correct maintenance and operating procedures can be enacted; for correct selection such as the colour code on electrical resistances; for marking of storage and transportation areas; and for quality purposes where it can seen at a glance the correct component has been used. Wimpey used colour in their timber frame houses for identifying fire stops and insulation. The colour of the decoration has an impact on producing a convivial working environment. Most offices use light pastel shades as a relaxing non-aggressive environment. Imagine working in a space with a purple ceiling, and the four walls coloured red, orange, green and black.
- *Noise*, or lack of noise, can be a contributory factor to lost production. People react differently to noise and those who generate it are usually less troubled than the recipients. The sound frequency bands and whether it is intermittent or continuous have different effects on the receiver. A loud unexpected noise, besides causing a problem at the time, can cause a level of uncertainty and hence loss of production if it is to occur again, but at an unknown time. Besides protecting the operatives working in a noisy environment with ear protectors, the source of the noise should be contained within the area if at all possible thereby reducing the impact on others.
- *Seating*, when correctly designed, can have a significant impact on output, but the solution will vary depending on the type of work. One of the best designs for work is the typist's chair. It can be raised or lowered, the back rest set to match the lumbar region and it can be set at a range of inclinations to suit the particular physique of the user. Without this the typist would suffer discomfort and eventual potential injury, both affecting output. Seats in lecture theatres must be such that

the student is not too comfortable so as to discourage sleep, but not too uncomfortable to restrict the process of leaning.

- *Amenities* can improve the working conditions and thus motivation, such as good toilets and general cleanliness.
- *Movements*, although not so relevant to many of the activities in the construction industry, the work study engineer is concerned with the movements of the operative, because if the movement of the operative can be eliminated or improved, productivity will also increase.
  - *Minimum movements.* Are the tools, materials, etc. close to hand and at an appropriate level so that it is not necessary to stretch, or to get up to collect?
  - *Simultaneous and symmetrical movements.* The aim is that the movement of the arms and hands should be balanced and used together if possible rather than one being idle. A good example in construction is the way a bricklayer lays bricks.
  - *Natural movements.* These are those that are naturally performed by the body such as lifting correctly by keeping a straight back and bending the knees. An example of unnatural movement is the act of tapping one's head whilst moving the other in a circular motion on one's stomach. Once this has been learned, try changing hands.
  - *Repetitive movement.* If these can be encouraged the body adapts and learns the movement and becomes more productive.
  - *Habitual movement.* If the body carries out a motion by habit as if on autopilot, the body becomes less tired and productivity increases. This means that all tools and materials should be in the same place so the operative does not have to think about it.
  - *Continuous movement.* Productivity is enhanced if the motions are smooth and not jerky. A good example is using a handsaw. If the joiner can use the full length of the blade the cutting will occur much easier and faster than if restricted to small movements.

Having taken all these factors into account where relevant, the engineer is now in a position to develop the new method and submit it for consideration by the management. This is crucial because it can only be installed if management approves it. There can be situations when the new method may be unacceptable, especially if it means eliminating too much labour that cannot be transferred elsewhere in the business, as that might cause a significant industrial relations problem. The report should state the benefits of the new method and include all the cost implications both in terms of saving and new costs incurred in setting it up. The development process results from careful examination of the facts as described in the examination process. To this is added the practical expertise of the work study engineer.

3.2.5 Install and maintain

Once the new method has been accepted it has to be installed. This can be a time-consuming process and requires the active co-operation of all the key players. This is not always easy to achieve especially if the changes are considerable, as demonstrated in the case study below. The method will only work if it is supported at the highest level of management. The senior management on the site must drive the installation forward and in this process, engaging and convincing the foreman, charge hands and operatives of the benefits that will accrue. If the workforce is composed of trade union members, their officials may have to be consulted, especially if it involves redundancies or affects their conditions of employment or service. To explain the consequences of installing the results of a new method and the impact of all concerned a case study is used.

Case Study 1

The pre-cast concrete wall manufacturing department is comprised of 4 batteries, each of which has 12 vertical cells. Each cell could make one wall unit sized approximately 2.4m high by 2.4m wide by 700mm thick and was separated from the next cell by a steel hollow dividing plate. The battery was so designed that after casting, hot air could be ducted under the battery, through the dividing plates and around the sides, enabling the concrete to reach a strength sufficient for removal after two hours after the casting process was completed.

An overhead gantry crane, concrete distributor and vibration equipment was dedicated to each pair of batteries, (1 and 2, 3 and 4). A gang of five operatives operated each pair of batteries.

Before the involvement of the work study engineer, the production from the wall shop was 3 batteries a day from the 4 batteries. However, the target production was to get 6 castings per day from these same 4 batteries and at the same time reduce each gang size from 5 to 4. Management saw this target

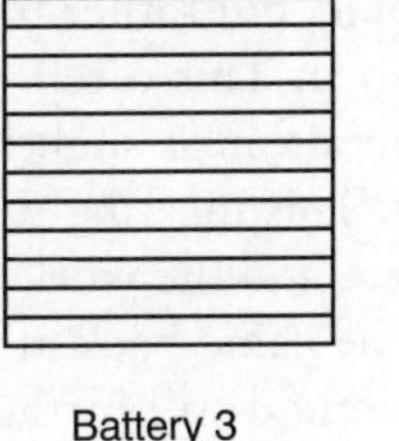
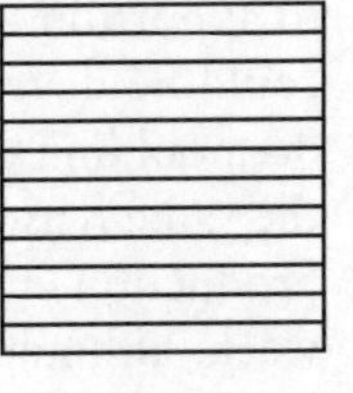
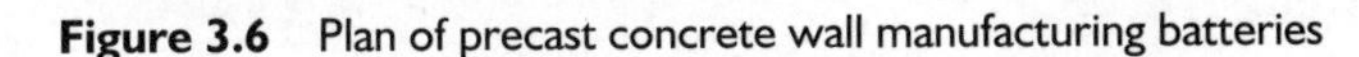

Figure 3.6 Plan of precast concrete wall manufacturing batteries

as achievable since the system used on the continent was regularly producing this level of output there, although the complexity of the components needed in the UK was greater.

The factory employees involved in the process were the production manager, who believed in the feasibility of the target, the foreman, the two gang leaders and the other gang members all of whom had been working very hard and conscientiously in obtaining the current production. When the production management advised the foreman of the targets, there was a significant scepticism about the viability of such a suggestion.

The proposed method was presented and discussed with the production manager, who called a meeting with the shop foreman, whose reaction had already been noted. He was asked not to comment on the accuracy of the timings and expected increases of production, but to see if the sequence recommended by the work study engineer was logical. After some practical minor amendments he agreed to follow the method to the letter. The proposition was then put before the operatives, who after 'muffled laughter', were content to follow the instructions. The two now redundant operatives were redeployed in the factory.

The work study engineer produced the proposed method using a multiple activity chart. This was drawn up on blackboards, one at each of the pairs of batteries, so that everybody could refer to it at any time. The time scale was vertical, the top of the board being the commencement of the shift. Each operative A, B, C and D (interchangeable in practice) was shown across the top of the board along with any equipment that might be required, as shown in Figure 3.7.

Although not shown here, activities were then drawn onto the chart demonstrating the start and finish times. On the first day progress was very slow and only achieved approximately 2 hours of the revised method. A meeting was called between the production manager, foreman, charge hands, work study engineer and maintenance engineer there in case it was necessary to make some small device to aid production and the work study engineer. After the 'I told you so' comments, a detailed analysis of each of the activities of the day's production was carried out. The meeting was not to apportion blame, but rather to establish causes of problems so solutions could be developed. This often meant making small tools to assist in the de-moulding operation or slightly modifying the concrete component for the same reason. This meeting process was carried out every evening after production ended, and the work study engineer withdrawing after the first week, although he came back at intervals to see if he could contribute to improving the method still further. It took nearly three months before the targets were achieved.

Time	Operator A	Operator B	Operator C	Operator D	Crane	Vibrator and chute
9.00						
10.00						
11.00						
12.00						
13.00						
14.00						
15.00						
16.00						
17.00						
18.00						

Figure 3.7 Multiple activity chart

The initial target was to complete the required output in about 10 hours. It is interesting to note that as this objective became more realisable, the operatives appeared to find this a challenge and decided they could improve on this and found alternative ways to improve the method. In the end they achieved the target in between 9 and 9.5 hours. It is suggested that a similar effect to the outcomes from Mayo's Hawthorne experiment happened because of the close involvement of the operatives in the decision making process (*Business Organisation for Construction*, Chapter 1).

Whilst it took three months to achieve the desired outcome, whenever a new method is introduced, it is not expected to reach the optimum outputs immediately. This is because even though the operatives are qualified in terms of skill and ability, they have to learn the process and this takes time. This process is referred to as the learning curve. The more the operation is repeated, the higher the productivity until meeting an optimum performance, as shown in Figure 3.8.

In general, the longer the length of the task and the more complex it is, then the longer it will take to learn. Also affecting the speed of learning is the capability of the worker and their motivation, how similar the task is to previous experiences, and the quality of instruction given.

Case Study 2

This example shows that changes to processes don't always come out as anticipated and can have unexpected results. In the same factory, floor components 1.2m wide by 200mm thick of varying lengths up to 6m were being manufactured on a conveyor system, each component taking on average 3 minutes to cast. The moulds located at ground level were constructed

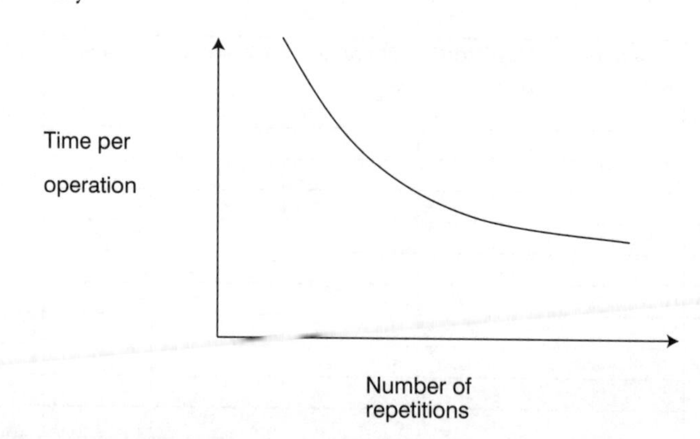

Figure 3.8 Learning curve

of steel. One of the processes, that of cleaning the mould ends, oiling and replacing them in position, was very tiring work as it was being conducted about 150mm off the ground and each component weighed approximately 65 kilograms.

It was decided that, although no productivity gain would be achieved but because of the tiring nature of the work, a compressed air hoist supported by an overhead gantry would be introduced to lift the mould ends onto a small table at a sensible working level. The table would be placed midway between stations A and B at either end of the mould, Figure 3.9. The mould end would be removed from the mould, lifted onto the table where it would be cleaned and oiled and then lifted back to the mould. Whilst this was happening the mould would be conveyed from station A to B.

The operatives were advised of the reasoning and instructed to carry out the task as the new method. After three weeks the foreman was told not to insist on the new method. Within a matter of days the operatives had resorted back to the original tiring method. On consultation the reasoning given was the new method, whilst less tiring, meant they had to work throughout the

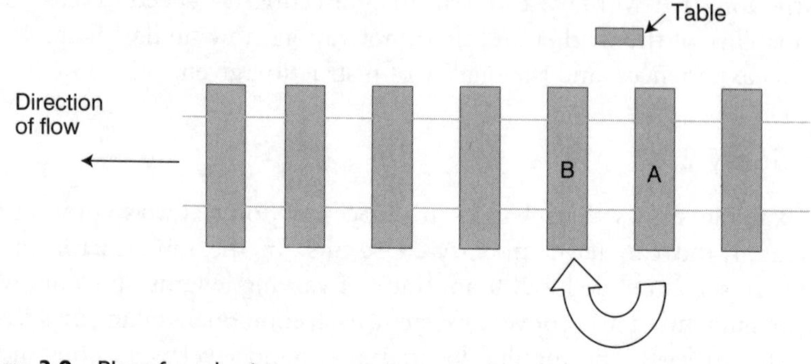

Figure 3.9 Plan of production

3–minute cycle. With the old method they could complete the task faster, have a short break before the next mould appeared in front of them. This was preferred.

3.3 Work measurement

3.3.1 Introduction

The scientific measurement of human work is necessary to establish how long different activities should take when a worker is engaged in that task at a reasonable expenditure of effort. It is not about working the operative to the point of exhaustion. The reasons for establishing these measurements are to evaluate past or present performance, or to predict future performance, as shown in Table 3.3.

3.3.2 Preparation for the study

Before beginning the recording of the work there are three issues to be addressed. First, that the job to be studied is being carried out in the most sensible way at the time, even though eventually the method may be changed. If there are obvious things wrong or the operatives have to carry materials long distances, climb over rubbish, etc., then this is clearly not the best available method.

Second, the operative must be qualified to carry out the task. Most jobs require an element of skill and if the person being studied is not experienced in the task being measured, then false readings are likely. Just because the bricklayer is a time-served craftsman, does not mean he is qualified, mainly because he may still be on the learning curve (see Figure 3.7).

Table 3.3 Uses of work measurement

Past and present performance	*Future performance*
Evaluation of alternative work methods	Producing schedules of work for planning purposes
Basis for fair incentive schemes	The preparation of estimates
Efficiency comparisons	The preparation of information for standard costing
	To be used as a basis for labour budgeting and budgetary control

Third, the job needs to be observed and then broken down into elements of short duration. These durations need to be long enough to be timed and recorded, but not too long otherwise the rate at which the operatives are working may change during the duration of the element

3.3.3 Time study or recording

The traditional means of recording is by observation with a stopwatch. It is used for repetitive work and semi-repetitive work. Where there are a repetitive and limited number of elements to measure in the complete task, there are hand-held electronic devices, which can be programmed, as shown in Figure 3.10.

These devices are excellent if all the elements of the operation can be determined before the study takes place, which is not always possible in construction work. The types of delays can also be programmed. Other facilities such as the rating at the time of study (section 3.3.4) are also included on the display. The engineer can then press the appropriate element, or non-productive activity panel, on the display and the device will store this information and the time that each activity started and finished. After the study the device is then connected to a computer when the information can be downloaded and analysed.

Where this kind of device cannot be used, a stopwatch is employed. These days these are digital, which has improved the accuracy of timing as the data can be held on the visual display whilst the engineer writes down the data, the timing still continuing unseen. There are two approaches, that of the 'fly back' timer and cumulative timing. The fly back restarts timing from zero each time the engineer commences timing the next element. In both cases each activity must be timed as frequently as possible so that on later inspection there are enough data to make a proper statistical analysis. Figures 3.11 and 3.12 illustrate typical work study sheets of fly back and cumulative timing respectively.

Element A	Element B	Element C
Element D	Element E	Element F
Element G	Element H	Element I
Element J	Element K	Element L
Element M	Element N	Non -Prod A
Element P	Element Q	Non -Prod B
Element S	Element T	Non -Prod C

Figure 3.10 Display panel

TIME STUDY SHEET		SHEET NO. 1
CONTRACT *Heywood Flats*	STUDY NUMBER	*1234*
OPERATION *Fix softwood linings to lift entrance*	STUDY REF	*CGM/421*
	DATE	*01/3/05*
	TIME STARTED	*08.15*
	TIME FINISHED	*12.18*
OPERATIVES *1 joiner*	TIME ELAPSED	*243 mins*
MACHINES	TOTAL O.T.	*241 mins*
OBSERVER *Charles Jones*	VARIATION	*0.82%*

CONDITIONS AT WORK PLACE	COMMENTS
Weather cold. Area of work artificially	
lit satisfactory. Access good	

ELEMENT	R	OT	BT	ELEMENT	R	OT	BT
Check time		*4.2*		*Check for alignment*	*100*	*1.9*	*1.90*
				with lift doors			
Pick up tool bag	*80*	*3.1*	*2.48*				
from box, walk 5m				*Lay down door frame,*			
to workplace				*mark out base to fit*	*95*	*2.0*	*1.90*
				over threshold bar	*100*	*2.0*	*2.00*
Take off plaster	*90*	*2.1*	*1.89*				
screeds				*Fetch trestle and*	*100*	*2.1*	*2.10*
				support base of frame			
Check frame for sq.	*95*	*2.0*	*1.90*	*ready to cut notch*			
ditto	*80*	*1.0*	*0.80*				
ditto	*85*	*1.7*	*1.45*	*Relax - fetch tea and*	*Rel*	*5.2*	*5.20*
				chat			
Take off bottom	*100*	*1.1*	*1.10*				
brace				*Saw along side of*			
				scribe with pad saw			
Rake out block joints				*(3.3m long)*	*110*	*2.0*	*2.20*
for plugs (4 no.)	*90*	*2.0*	*1.80*	*ditto*	*105*	*2.0*	*2.10*
ditto	*100*	*2.0*	*2.00*	*ditto*	*100*	*0.9*	*0.90*
ditto	*90*	*1.8*	*1.62*				
				Ditto other side of			
Cut two plugs	*I.T*	*1.9*	*1.90*	*scribe*	*100*	*2.0*	*2.00*
Split - timber us				*ditto*	*95*	*2.0*	*1.90*
				ditto	*100*	*1.7*	*1.70*
Cut four plugs (soft							
wood)	*100*	*2.0*	*2.00*	**Continued to the**			
ditto	*110*	*1.2*	*1.32*	**last sheet of the study**			
P/U frame, carry							
4 paces to opening	*105*	*2.0*	*2.10*	*Check time*		*6.2*	
Try fir in opening	*100*	*2.0*	*2.00*				
ditto	*80*	*2.7*	*2.16*				

Figure 3.11 Time study sheet – fly back time (adapted from Geary (1962) with permission)

TIME STUDY SHEET		SHEET NO 1
CONTRACT	*Bacup distillery*	
OPERATION	*Place and finish concrete to tank base type 4 A (drawing ref BD23/4)*	
STUDY NO	*341*	
REFERENCE	*CGM.206*	
DATE	*10/2/04*	
TIME STARTED	*09.20*	
TIME FINISHED	*12.05*	
OPERATIVES	*3 labourers & 1 ganger*	
TIME ELAPSED	*165 mins*	
MACHINES	*10 tonne mobile crane, 0.75m conc. skip*	
TOTAL O.T.	*165 mins*	
OBSERVER	*Peter Williams*	
VARIATION	*none*	
CONDITIONS AT WORK PLACE	*Weather cold 42°C. Light breeze Dull. Occasional light rain*	
COMMENTS	*Working conditions difficult due to the amount of steel reinforcement. Banksman unable to assist due to position*	

ELEMENT	R	OT	BT	ELEMENT	R	OT	BT
A - ganger							
B, C, D - labourers				*A, B, C, D - clean out*	*90*	6.0	5.4
				base			
Check time		*8.45*				*0.25*	
		4.5					
		49.5		*A, B, C, D - ditto*	*100*	6.0	6.0
A, B, C, D - Fetch tools	*90*	6.0	5.4				
Etc. from tool box						*0.40*	
and walk 20m to				*A, B - start up*	*95*	4.0	3.8
tank base				*vibrator*			
		51.0					
				C, D - swing in full	*75*	4.0	3.0
A, B - pick up poker	*90*	4.0	3.6	*Concrete skip*			
vibrator from						*06.0*	
adjacent bay				*A, B - lay boards*	*110*	4.0	4.4
				Over reinforcement			
C, D Fetch vibratory	*100*	4.0	4.0	*and support*			
screed and position							
along side base				*C, D - as before*	*70*	4.0	2.8
		53.0				*0.80*	
				A, B - as before	*100*	3.0	3.0
A - talk to foreman	*100*	5.5	5.5	*C, D - as before (skip*	*75*	3.0	2.3
				away)			
B, C, D - wait	*I.T*	16.5				*0.95*	
				A, B - as before	*95*	2.0	1.9
		58.5					
				C, D - wait	*I.T*		
A, B, C, D - clean out	*100*	10.0	10.0			*10.5*	
base				*A - vibrate*	*100*	2.0	2.0
	9/	*01.1*		*B - spread conc.*	*110*	2.0	2.2
				with shovel			
				C, D - swing in full	*100*	4.0	4.0
				Skip and discharge			
						12.5	
				Etc.			

Figure 3.12 Time study sheet – cumulative time (adapted from Geary (1962) with permission)

The stopwatch is started, perhaps in the office or as the engineer arrives at the place of work, and when the first element starts the watch is used to record this point in time. The time up until this point in time is referred to as the check time and is used to allow the engineers the opportunity to organise themselves, perhaps recording the weather conditions, making notes about the working conditions and so on. The same occurs at the end of the study, so after the final element the engineer can at their leisure end the study. These check times are entered in the observed time (O.T.) column. As is demonstrated in the figures descriptions of the various elements are entered as they occur along with the observed times. Even on this small sample it can be seen that the electronic devices outlined above would have inadequate spaces to pre-record all the activities.

Non-productive work, such as the cutting of two plugs which split because the timber was 'useless', are entered as idle time (I.T.) and when the operative has a tea break this is considered as a relaxation allowance (Rel). In Figure 3.12 the foreman talks to the ganger. This is always rated at 100, as it would be difficult to assess the effectiveness of the communication (section 3.3.4).

3.3.4 Rating

Timing the length an element takes is not enough in itself. The operative can appear to be working very hard or alternatively very slowly. For example, an operative may be loading the shovel with a small amount of earth and doing this rapidly, or filling the shovel and doing it much slower. The latter may be more productive if one measures the amount of soil shifted over a period of time. This is where the experience of the work study engineer comes into play. The engineer has to assess the rate at which the operative(s) are working against a standard norm. This is referred to as rating and it is entered in column R in Figures 3.11 and 3.12. There are various rating scales but the one used in the UK is where the standard norm is 100. To give some indication of this amount of effort, it is said that an average sized man used to walking achieves a standard rating of 100 when walking at 4.63 kilometres per hour (4 miles per hour) on level ground, taking reasonable rest periods and not carrying any load.

BS 3138 gives some further descriptive comparisons of rating in five-point graduations such as:

- 125 – Very quick; high skill; highly motivated
- 100 – Brisk; qualified skill; motivated
- 75 – Not fast; average skill; disinterested

The observed times have to be normalised as if the activities recorded and timed are carried out at the standard rating of 100. This is referred to as the basic time (B.T.) as shown on the time study sheets, Figures 3.11 and 3.12. This is so that when the data are eventually used, one is comparing like with like. The calculation is:

$$\text{Basic time (B.T.)} = \frac{\text{Observed time (O.T.)} \times \text{The rating (R)}}{\text{Standard Rating (100)}}$$

In this case the standard rating is the norm of 100. So an observed time of 0.87 minutes given an observed rating of 95 gives a basic time of 0.83 minutes. In other words, if an operative is working below par, the basic time will be lower and if working harder than the norm, the basic time will be higher.

In the case of the fly back timed study example, Figure 3.11, the observed time is as recorded, whereas in the cumulatively timed study, Figure 3.12, the observed time is calculated based on the number of operatives working on any activity. For example, the four operatives A, B, C and D commence fetching their tools, etc. at 08.49.5 and complete at 08.51, a total of 1.5 minutes, but since there are four of them, the total observed time for that activity is 6 minutes and with an observed rating of 90, the total basic time is 5.4 minutes. When the study is completed the information is abstracted and entered onto the summary sheet, Figure 3.13. The descriptions of similar elements may vary on the time sheet at the time of recording, but on analysis turn out to be the same. These are entered on the summary sheet along with the total basic minutes spent carrying out the element, the quantity of work done and the method of measure, such as job, metre run, squared or cubed.

3.3.5 Relaxation factors and relaxation times

It is unrealistic to expect operatives to work without a break and equally to ignore the effort and posture required to carry out the task and the energy expended. Relaxation factors and times are applied from various categories such as energy output, posture, motions, personal needs, and thermal and atmospheric conditions. Some of these can be considerable and wide ranging. For example, energy output can be as little as less than 5 percent for sedentary work to 50 per cent for very heavy work; posture can vary between 0 and 10 per cent; motions up to 15 per cent when working in very confined conditions, extremes of temperature and very dusty conditions (up to 20 per cent). Personal needs such as going to the toilet, having a cup of tea, etc. is added to every activity, as it is not practical to isolate this to a specific element. This is 8 per cent for a man and 10 per cent for a

SUMMARY SHEET

OPERATION	
Place and finish concrete to reinforced concrete tank base using mobile crane and 0.75m skip	STUDY NO *341* REFERENCE *CGM/206* DATE *10/2/05*

3m

4m

plan

0.750m

section

ACTIVITIES	Total B.T'S	% RELAXATION A	B	C	D	E	F	G	% Con	% Add	Total SMs	Quant	Unit Std Times
Fetch tool and prepare	*40.2*	*5*	*1*	*1*	*1*	*8*	*-*	*2*	*5*	*23*	*49.4*	*1 job*	*47.4/job*
Clean out base in preparation	*21.4*	*8*	*3*	*2*	*2*	*8*	*-*	*2*	*5*	*30*	*27.8*	*12m²*	*2.2/m²*
Swing in and empty skip	*88.9*	*14*	*2*	*3*	*2*	*8*	*-*	*2*	*5*	*36*	*120.9*	*12 no*	*9.7/no*
Spread concrete	*81.1*	*25*	*5*	*3*	*2*	*8*	*-*	*2*	*5*	*50*	*121.7*	*9m³*	*14.2/m³*
Vibrate concrete	*38.8*	*10*	*5*	*2*	*1*	*8*	*-*	*2.5*	*5*	*33*	*51.6*	*9m³*	*5.5/m³*
Level and tamp concrete	*57.7*	*15*	*5*	*3*	*2*	*8*	*-*	*2*	*5*	*40*	*80.8*	*12m²*	*6.4/m²*
Float off surface of concrete	*166.3*	*10*	*5*	*4*	*2*	*8*	*-*	*2*	*5*	*35*	*249.5*	*12m²*	*18.0/m²*
Clean surplus concrete from formwork and remove concrete	*20.1*	*8*	*4*	*3*	*2*	*8*	*-*	*2*	*5*	*32*	*26.5*	*14m*	*1.8/m*
Wash tools and clean up	*32.8*	*3*	*3*	*2*	*2*	*8*	*-*	*2*	*5*	*25*	*41.0*	*1 job*	*39.4/job*
Ganger's supervision.	*38.8*	*6*	*2*	*1*	*2*	*8*	*-*	*2*	*5*	*28*	*49.7*	*1 job*	*47.7/job*
Total	*586.1*										*818.9*		

Figure 3.13 Summary sheet. (adapted from Geary (1962) with permission)

woman. A more complete breakdown can be found in Currie's *Work Study*, 4th edition. In construction, the total of these can be in excess of 50 per cent. The relaxation columns in the summary sheet, Figure 3.13, are filled in with the relaxation times A–G representing different categories of relaxation.

It is normal to add up to 5 per cent contingency to each of the elements to take account of variances and inconsistencies that might have occurred during the study. These percentages are then totalled and added to the basic times giving the standard minutes. At this point there is a scientific and rationalised number that can be used to state how much time the given quantity of work takes in this particular situation.

3.3.5 Synthetics

Each study has error built into it and is particular to the specific conditions of the operation when and where the study was taken. The decision to select a rating and relaxation factor is another source of potential error. Therefore to take and use the standard minutes from one study could be misleading, when applied to similar work scenarios. However, if similar data from several other studies are brought together, a mean can be calculated and the accuracy of the data improved. This can be used with a degree of certainty for planning, incentive schemes, costing schemes and estimating purposes. This information is stored in what is called a synthetic library. The library takes a considerable time to accumulate into a useful source of information, and can be costly to produce. The procedure for establishing synthetic data is as follows:

- At least 10 studies need to have been completed on similar operations to have enough data available for a sensible statistical analysis.
- Basic times of similar activities along with the quantity of work done are abstracted from the work study sheets and entered onto a time study abstract sheet.
- From this an analysis can be carried out to establish a basic time that reflects the true situation. There can be readings in the data that are clearly not typical and these are excluded. The more recordings that have been made the more accurate the calculation of the final basic time.
- A percentage allowance for the relaxation factor and time can be added along with a contingency factor to produce the standard minutes.
- This is then placed in the library, either as a new entry or used to modify existing entries.

Besides using this information for estimating, planning, etc., it can be used to establish how long an operation, never performed before, will take. The operation can be analysed and broken down into its elements and synthetic data then applied. When totalled up and an appropriate contingency percentage added, depending on the amount of uncertainty, a reasonable estimate of its overall duration can be predicted.

3.4 Activity sampling

The previous discourse dealt with the time or sequence of activities. Activity sampling is concerned with establishing the frequency at which operatives are working. It does not measure the effectiveness or efficiency of the work being carried out. It is ideal for the manager to obtain a feel for what is happening in the workplace, thereby giving an indication of the overall health of the operation. The method can also be used for establishing the usage of rooms and plant.

The process is relatively simple. The observer travels the whole area of the workplace and records activity, or the lack of it, ensuring that the same person is not recorded twice. This is known as the field count. It is important to travel the workplace several times, at random and in different directions. The frequency and randomness determine the accuracy of the work. For example, if sampling room use, looking at the use of classrooms only at the time between lessons would clearly give a false reading about their usage. When taking an activity sample of operatives, whilst it is not expected to count and observe everybody working on the site, it is usual to have seen at least 75 per cent of those who should be in place, otherwise the reliability of the study becomes suspect.

The observer counts the number of operatives that are active and those classed as non-active. Those defined as active are operatives and supervisors that are:

- Physically engaged in carrying out their trade or and activity that is related to it.
- Assisting others doing work such as holding the end of a timber whilst another joiner is cutting it.
- Driving or operating of a vehicle or machine, or assisting in it such as acting as a banksman to a crane or excavator.
- Carrying materials from one place to another or moving a load.
- Giving or receiving instructions.
- Watching over or standing by equipment or machines for safety or control reasons. The plant may be wholly or part automatic but the operative has to be there in case human intervention is required.

As can be seen from these descriptions it takes no account whether the work being carried out is the most effective or efficient way of carrying out the task.

Inactive observations are:

- waiting for others to complete an operation
- waiting for materials to arrive
- waiting for instructions to be given
- standing or travelling with for no apparent productive reason
- riding on vehicles when not required.

The count is conducted using two counters, one for the total number observed and the other for those classed as active. Figure 3.14 is an example of a field study count.

Active percentage is expressed as a percentage of those observed as active as against the total number observed. In the example shown, the numbers present increased after lunch, as the result of two persons arriving late. Those present can be checked at a later stage either from clock cards or the reporting-in book. The comments column is useful as it can indicate any mitigating factor affecting the observations, for instance, heavy rain or high wind stopping the tower crane from operating. If it is observed that the activity appears especially low in a particular part of the site then more detailed studies can be undertaken.

In the example shown the column 'Active' varies from 53 to 60 per cent, but this is not surprising when reminded of the fact that relaxation factors can be up to 50 per cent in construction work. However, if the definitions of active and non-active are revisited it can be seen that these are not measurements of productivity and non-productivity. Many of the active observations could be inefficient, which demonstrates there are major possibilities for improving productivity, by investigating methods of work.

It is seen in the field count shown in Figure 3.14, the percentage of active workers varies throughout the day. This is not unexpected, as it has been found that, generally, there is a similar changing pattern of work activity in most work scenarios. These variances are demonstrated in Figure 3.15.

On reflection this is not surprising. Most people at work gradually build up self-motivation as the morning progresses, but as lunch-time approaches begin to turn their mind to the break. After lunch there is a rapid increase to one's peak followed by a gradual decline to the end of the working day. The majority of people fall into this pattern, although there are exceptions. A supervisor should be aware of this trend and place

FIELD STUDY ACTIVITY COUNT							CONTRACT: *Jasper Winston House*
							REF NO: *CGM/56*
							STUDY NO: *435*
Date	Time of count	No present	Observed No	Observed %	Active No	Active %	Comments
02/03/05	*8.25-8.34*	*90*	*80*	*89*	*44*	*55*	*Sunny*
	8.54-9.07	*90*	*81*	*90*	*45*	*56*	*Sunny*
	9.13-9.24	*90*	*82*	*91*	*47*	*57*	*Sunny*
	10.03-10.16	*90*	*81*	*90*	*47*	*58*	*Clouding over*
	10.40-10.53	*90*	*80*	*89*	*48*	*60*	*Clouding over*
	11.07-11.19	*90*	*81*	*90*	*49*	*60*	*Clouding over*
	11.30-11-41	*90*	*83*	*92*	*49*	*59*	*Cloudy*
	11.50-12.03	*90*	*82*	*91*	*47*	*57*	*Cloudy*
							Lunch break 12.30-1.00
	13.10-13.22	*92*	*84*	*91*	*46*	*55*	*Cloudy*
	13.45-13.56	*92*	*83*	*90*	*47*	*57*	*Occasional light rain*
	14.15-14.27	*92*	*83*	*90*	*49*	*59*	*Occasional light rain*
	14.38-14.52	*92*	*85*	*92*	*51*	*60*	*Occasional light rain*
	15.07-15.19	*92*	*86*	*93*	*53*	*59*	*Cloudy*
	15.30-15.41	*92*	*84*	*91*	*48*	*57*	*Cloudy*
	16.00-16.11	*92*	*83*	*90*	*46*	*55*	*Cloudy*
	16.22-16.33	*92*	*82*	*89*	*44*	*53*	*Cloudy*

Figure 3.14 Field study activity count

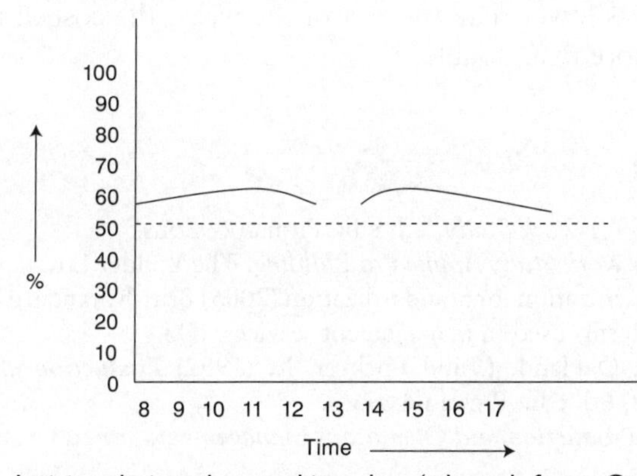

Figure 3.15 Activity during the working day (adapted from Geary (1962) with permission)

emphasis on motivation operatives at the lower activity times, bearing in mind of course, the supervisor will probably be of the same attitude.

If the same sampling approach is applied to the full working week, it is noted that there is a variance of activity throughout, as demonstrated in Figure 3.16.

Again, reflecting on the attitude of most people, one comes to work suffering the ‘Monday morning blues’, but as the day and week progresses motivation increases until Friday when the thought of the weekend takes

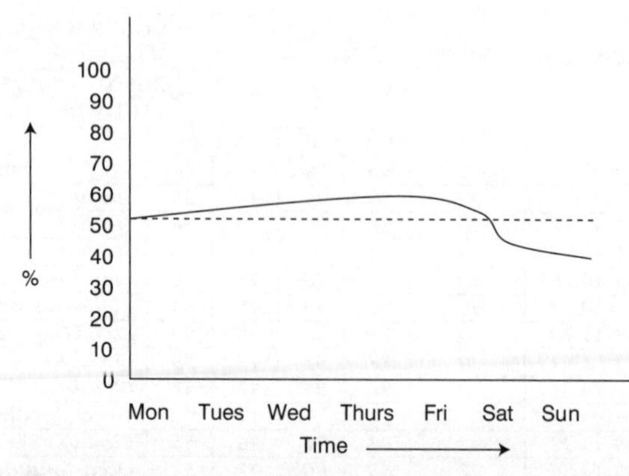

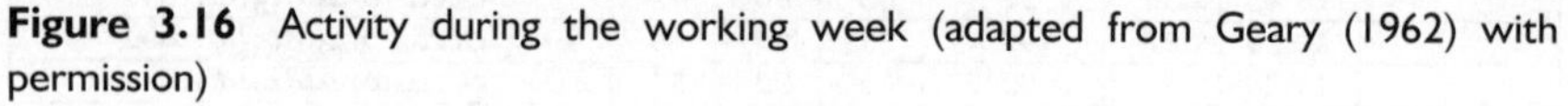

Figure 3.16 Activity during the working week (adapted from Geary (1962) with permission)

over and motivation decreases. If asked to work on Saturday, motivation is much reduced, not helped by the fact that people tend to come casually dressed, and Sunday working the activity rate is generally even lower. This highlights another issue that on Sunday, operatives are paid double time and if their activity is lower than the rest of the week, the cost of the output is considerably more than double.

References

Currie, R.M. (1997) *Work Study*, 4th edn. Pitman & Sons.

Geary, R. (1962) *Work Study Applied to Building*. The Builder Ltd.

International Organization for Standardization (2005) British Standard BS 3138:1992 Glossary of terms used in management services. IOS.

Muhlemann, A., Oakland, J. and Lockyer, K. (1992) *Production and Operations Management*, 6th edn. Pitman & Sons.

Wild, R. (1995) *Production and Operations Management*, 5th edn. Cassell.

CHAPTER

4

Health and safety

4.1 Introduction

There are four good reasons why safety should be taken seriously. First, the legal obligations which, if flouted, the offender could see a jail sentence or significant fine; second, whenever anybody is injured, irrespective of severity, it costs the company money; third, all of us, employees and employers, have a moral obligation to look after each other; and fourth, by doing so you increase the likelihood of not having an accident oneself.

The construction industry has one of the highest accident rates of all industries. In 2006/07 there were 3.7 fatal injuries per 100,000 employees, or 77 deaths, and 32 per cent of all worker deaths were in the construction industry. There were over 3,700 reported major injuries with the highest being slip and trip (27 per cent) and falls from a height (also 27 per cent). The number of over-three-day injuries was reported to be over 7,100; the most common cause (38 per cent) involved handling, lifting or carrying. It is unacceptable to talk about accidents as being an occupational hazard; the industry has to, and is beginning to, reduce the accident figures, and is aiming to achieve an incident- and injury-free place of work.

4.2 Definitions

The terms *risk* and *hazard* are commonly used terms when dealing with safety and are regularly misused, so at the outset the reader should understand the difference. A hazard is something with potential to cause harm. The effect can cause a physical injury or impact on one's health. A risk is the likelihood of that potential being realised.

4.3 Legal obligations – background

There are numerous pieces of legislation concerned with health and safety, but the key one is the Health and Safety at Work, etc. Act 1974 (HSW), and it is within this act that there is the potential for managers to be sent to jail for negligence. Currently, this seldom happens, but there has been increasing pressure for this and/or corporate manslaughter charges to be brought to those who flout the law. This is in response to the various rail tragedies of the last few years and the *Marchioness* accident on the river Thames in 1989. The Health and Safety Executive (HSE) is also pressing for one senior person in the company to be responsible for all matters concerning safety, but employers are resisting this, no doubt preferring to hide behind the shield of corporate responsibility, which makes it more difficult to apportion blame. A more detailed look at the implications of the HSW and other key regulations are considered in section 4.9.

The UK has to comply with the European Union legislation though acts of parliament and enabling acts. This is because under the Treaty of Rome (1957), Article 118A, member countries have to 'pay particular attention to encouraging improvements, especially in the working environment, as regards the health and safety of workers'. Directive 89/391/EEC was produced as a result of this and was enacted in the UK through the Management of Health and Safety at Work Regulations 1992, revised 1999.

Other Directives followed:

- Directive 89/654/EEC: the workplace directive
- Directive 89/665/EEC: the use of work equipment directive
- Directive 89/656/EEC: the personal protective equipment directive
- Directive 90/269/EEC: the manual handling of loads directive
- Directive 90/270/EEC: the display screen directive.

These have been enacted in UK legislation as:

- The Workplace (Health, Safety and Welfare) Regulations 1992
- The Provision and Use of Work Equipment Regulations 1992
- The Personal Protective Equipment at Work Regulations 1992
- The Manual Handling Operations Regulations 1992
- The Health and Safety (Display Screen Equipment) Regulations 1992.

A further directive of special interest to the construction industry was produced entitled Directive 92/57/EEC: the construction sites directive. This is now implemented in the UK as the Construction (Design and Management Regulations) 2007.

There are important further regulations that have a direct bearing on the construction industry. These are:

- The Reporting of Injuries, Diseases and Dangerous Occurrences Regulations 1995 (RIDDOR)
- The Construction (Health, Safety and Welfare) Regulations 1996
- The Control of Substances Hazardous to Health (Amendment) Regulations 2002 (COSHH)

4.4 Financial costs of an accident

Every accident costs money. Even a small injury resulting in the need for minor treatment, such as the application of a plaster, results in the operative stopping work with loss production as well as the first aid worker's time. It has been suggested that all these minor incidents annually equate to the cost of building one mile of new motorway. If it is necessary to go to the an accident and emergency department, the costs rise and the operative could be off work. A fatality can cost many tens of thousands of pounds.

In the UK, 30 million days a year are lost from work-related injuries and ill health valued at £700 million. On the assumption that construction represents some 7 per cent of gross national product (GNP) and it is on average twice as dangerous as the majority of other industries, then as a very rough indication the cost to the construction industry is approaching £100 million per annum. Looking at it another way, if the estimated turnover of the industry is £50 billion then this is the equivalent of 2 per cent turnover. Since, in the year 2002/03, 31 per cent of all industrial accidents in the UK occurred in the construction industry, the figure could be even higher.

Fatal accidents can be very costly to the company depending on the amount of negligence apportioned to the management of the organisation. Other costs include:

- The loss of production resulting from the trauma and hence the overall morale of employees.
- A cessation of production due to an enforcement order being placed by the Health and Safety Executive if they are not satisfied that the work situation has been made safe or they need to conduct further investigation.
- Costs of staff involved in sorting out the accident and further investigation costs.
- Costs involved in attending inquests and other meetings associated with police, coroner, trade unions, insurance companies, and one's own safety department.

- Short-term payouts, without prejudice, to relatives to assist them through the immediate aftermath of the accident.
- Compensation awarded by the courts. Whilst this might be covered by insurance the premiums may well rise the following year as a result.
- Any damage caused to property as a result of the accident.
- The costs involved in training a new member of the team.
- Reduction in the company's reputation.

In the case of incidents where the operative is off for three days or more they will have to be replaced, or the gang will continue to work inefficiently. The replacement member of the gang has a learning curve and there is an administrative load to manage this change.

Another cost, which does not affect the company directly, is the financial burden placed on the deceased relatives through the loss of income, which can take several years to be awarded. Because of this delay, the Lighthouse Club Benevolent Fund was established in the 1960s. This charity gives aid and assistance to construction workers who suffer an accident or ill health, and to their families during this waiting period.

Finally, there are the intangible costs: the loss of a family member and colleague can have an untold emotional impact.

4.5 Moral obligations

There is clearly a moral responsibility on managers and companies to look after the welfare of those employed. To place someone into an unsafe environment is unacceptable and yet this is done and not always unwittingly. In the post-war years it was still commonplace to talk about industrial diseases and accidents as occupational hazards. This changed as legislation was brought in to protect the workforce. It would now be unacceptable to consciously put workers into a known hazardous work situation unless it was unavoidable and sensible protective measures had been adopted, for example, during the removal of asbestos.

There is, however, still one notable exception – stress – although even this is now taken more seriously and steps are being taken to reduce it (*Business Organisation for Construction*, Chapter 9). We continue to allow employees to be subjected to intolerable stress conditions as a result of long hours, pressure of work, harassment and, sometimes, the employee's own choice. The result is lost time through illness, domestic unrest, reduced quality of output, or in extreme cases, burnout. In certain work environments, working too hard it is thought to be macho, so peer pressure plays its part.

4.6 Self-preservation

If safety is taken seriously, actions will be put in place to make the workplace safer and to improve attitudes to health and safety and the risk to everyone will eventually decrease. However, the construction site remains a potentially dangerous place and there is concern that those who are not 'street wise' can place themselves in greater danger. Two examples follow.

Incident 1

Whilst working on a sheet roof with two experienced operatives, a young trainee fell through the roof to his death. The client had approved the sub-contracting firm's method of safe work. This involved the use of cat ladders on the roof with clear instructions that no work was to be carried out except from these ladders. It is believed that the two experienced workers walked along the roof from one cat ladder to the next avoiding the need to climb down to ground level and re-ascend at the next ladder. They walked along the bolts securing the roof sheeting to the purlin, which supported their weight. The trainee on the other hand walked on the unsupported roof.

Incident 2

In Hong Kong it is common practice to use hand-bored caisson piles. The operative carrying out the excavation, excavates a circular hole some half a metre deep and then using a metal former, concretes a ring around the perimeter of the excavation. This process continues to the required depth, which can be up to 30 metres below ground level. A hoist is positioned over the top of the excavation to lift out excavated materials and to lower the concrete for the sides of the caisson. A pump and compressed-air line are also lowered into the excavation, the latter to reduce the risk of methane gas getting into the working area. These are often husband and wife teams, the man working in the excavation and the wife operating the hoist.

During one of these excavations, a scream was heard and a young UK engineer rushed to the scene, and seeing the operative unconscious at the base of the excavation, instructed the hoist operator to lower him down to assist. He was overcome by fumes and died. A second young engineer repeated his colleague's ill-advised actions and also perished. An experienced person would have realised the probable cause of the unconscious operative's plight, stood back, rationalised the situation, and if breathing apparatus and a qualified user were not available, would have called the fire brigade.

The final goal is to get to a point where safety awareness is part of everyone's daily life, not just at work, but in also in other situations. So, for example, using the correct fuse in domestic appliances, checking the polarity of sockets, closing doors at night to stop the spread of smoke and fire, installing smoke detectors should be the norm. If an employee visits a site and observes an unsafe practice, action can be taken by reporting it to the appropriate supervisor or taking direct action, if appropriate. Visitors, however, can only bring this practice to the attention of the site management, but should be encouraged to do so and not made to feel that they are interfering.

Direct action can be taken by all to ensure that any, even minor, indiscretions are dealt with, such as nails protruding from wood or loose reinforcement tie wire, and the perpetrators educated not to do it in the first place.

4.7 The impact of an accident on others

It should be remembered when a serious accident occurs, work colleagues and family members have to deal with the loss and to demonstrate this, the scenario poses the question 'what would you feel like in that position?'

Scenario

In the incident in question an operative was out of sight underneath a machine making a modification whilst the rest of the gang were having their tea break. This was a daily event each morning. He had been instructed to take the ignition key to the machine with him so that the machine could not be started inadvertently whilst he was still underneath. On the day in question he had not done this. After the tea break the foreman instructed the charge hand to start up the machine, and the operative underneath was impaled though the chest with two 100mm diameter steel tubes driven in with a 50 horse power electric motor. He died within a few minutes after having been extracted from the situation.

Questions:

- Even though it was not his fault, what would you feel if you were the charge hand and had started the machine up?
- How would you feel if you were the foreman who gave the instruction?
- How would you feel as one of the other members of the gang, since they had all worked together in excess of three years?
- What goes through the mind of the first-aid worker underneath the machine trying to keep the injured man alive?

- How does the manager feel and react, while having to take charge of the situation until the emergency services come, and after they leave continue his command?
- What goes through the mind of the person going to the injured man's home to tell his wife that her husband has been injured and then escorts her to the hospital?

The point of this example is it shows how far-reaching the impact of an accident goes. It is not just the loss of a loved one and his income for the family. The incident does not just stop after the day of the accident. As the questions imply, the trauma for those involved continues, not helped that they will be interviewed several times more over the next couple of years whilst a case is being prepared either for court or an insurance claim for compensation. Further, it is not hard to imagine as a result of the trauma, it takes time before production reaches its pre-accident rate.

4.8 What is the problem? A statistical analysis

Frank Bird produced the information shown in Figure 4.1 indicating the relationships between the different types of event. For every fatal or serious accident there are 10 minor injuries, 30 incidents of property damage (in the case of construction, the building, materials, components or plant) and 600 'no-injury accidents'. So, for example, it takes on average 600 bricks to be dropped off the top of the scaffolding before a person is eventually hit and seriously injured or killed.

Delving deeper, by looking at the statistics of accidents, it is possible to discover the number of accidents, the types and causes of accidents and trends, and compare these to other industries. The prognosis is not good.

The main source of statistical data is the HSE which publishes annual health and safety statistics. These documents provide information going back over the previous 10 years showing the accidents occurring in all workplaces. This means that comparisons can be made between different industrial sectors and can help spot trends in the construction industry. The comparisons are based on the number of accidents per 100,000 employees working in an industry or sector. All the statistical data in this section are sourced from various HSE publications.

The data, as presented, should not be considered in isolation, as there are other issues that need to be analysed. For example, just because accident trends are going in a particular direction, does not mean that the safety of the industry is getting better or worse. It also depends on the annual volume of turnover and the type of work being carried out. The greatest

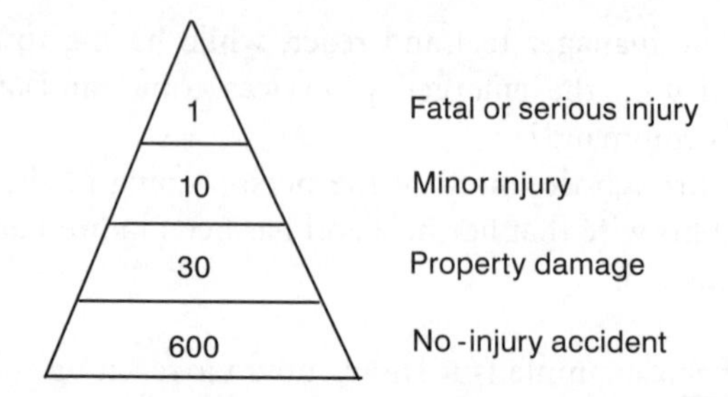

Figure 4.1 The accident pyramid showing relationships of accidents

causes of death in construction are from falling persons or falling objects landing on people. It might be argued that the greater the percentage of the turnover comprising tall buildings, the more likely there will be an increased number of fatalities. Alternatively tall buildings are more likely to be built by the more safety-conscious contractors and, therefore, the numbers injured working on these buildings is fewer.

Other variables include: whether there is any difference between the larger contractors' safety records and the those of smaller operators; if some trades are more susceptible to accidents than others; does the age and experience of the operators affect the likelihood of an accident; and are there any regional differences in accident rates. If there are any differences to these questions, then an analysis as to why needs to be conducted.

Statistics are based on information provided as a result of the Reporting of Injuries, Diseases and Dangerous Occurrences Regulations 1995 (RIDDOR) in which the classifications used are: fatal and serious accidents, those that cause at least three days off work, and diseases and dangerous occurrences (section 4.9.2).

From the statistical information trends in the industry can be established. These statistics vary from year to year and it is suggested if the reader needs the most up-to-date data, they consult the HSE web site. The rate of fatal injury row is calculated on the number of fatalities per 100,000 workers and is significant as it more clearly demonstrates the trend in the safety performance of the industry rather then the numbers killed. Similar data are collected for reported major injuries and over-three-day injuries. By transposing the rates of fatalities, major and over-three-day accidents onto the same graph, the ideas proposed by Frank Bird are confirmed that there is a relationship between the numbers and types of accident.

The HSE breaks down these data further into the causes of fatal, major and over-three-day accidents and identifies that over the years, the main causes of fatal accidents on construction sites are falls from heights (over 30 per cent), being struck by an object or moving vehicle, contact with electricity and then trapped due to collapse of excavations.

In the case of major accidents, there are similar causes but accidents caused by handling and lifting, slips, trips, and falls become more significant. This is reflected in over-three-day accidents.

Each of the categories of types of accidents can be broken down into the causes of accidents and this shows management where to concentrate improvement efforts. This is necessary, as whilst the aim should be to eliminate all accidents, this does not happen overnight, so it is sensible to concentrate on the areas of greatest risk. See Pareto's Law (*Finance and Control for Construction*, Chapter 13).

Tables 4.1 and 4.2 are concerned with fatal accidents associated with maintenance. Table 4.1 shows the main types of accidents which can cause fatalities, and the approximate percentage of each.

The two most serious causes of fatalities are falls and being crushed or entangled in machinery and plant and it would be sensible to further investigate the causes of these first. Falls is a broad description and as shown in Table 4.2, there are many different causes, some of which are more frequent than others, such as falling off ladders, roofs and through fragile

Table 4.1 Causes of fatal accidents in maintenance work

Type of accident	*Percentage*
Falls	30
Crushed or entangled	24
Asphyxiation	9
Electrocution	7
Burns	6
Impact	3
Falling objects	21

Table 4.2 Types of falls during maintenance

Type of accident	*Percentage*	*Type of accident*	*Percentage*
Flat /sloping roofs	22	Plant	6
Ladders	22	Scaffolding collapse	4
Fragile roofs	22	Window cills and ledges	3
Work platforms	11	Cranes and gantries	1
Scaffolding	8	Others	1

roofs. Each of these can be broken down further. For example, the causes of falls from ladders could be as a result of not tying the ladder properly at the top, insufficient gradient, or slipping whilst ascending or descending. From this detailed analysis, steps can be put in place to reduce the risks of these occurrences. The process can be applied to other types of accidents.

A further investigation can reveal who is responsible for the cause of the accident. Blame can be apportioned to that of the management, workers, a combination of both, third parties or unknown. What is found to be significant is that it is not unusual to find that management is responsible in full or in part for nearly 70 per cent of all accidents associated with maintenance.

4.9 Key legislation and regulations for the construction industry

Whilst not a complete summary of all legislation, those summarised in this section are the ones most commonly cited and referred to.

4.9.1 Health and Safety at Work, etc. Act 1974

This is an extensive piece of legislation so only the main elements of the general duties and enforcement are briefly mentioned.

Section 2 – General duties of the employer to employees

It is the duty of every employer to ensure, 'so far as is reasonably practicable', the health, safety and welfare of their employees. In particular this includes:

- That the provision of plant and equipment, its maintenance, and systems of work are safe and without risks to health. On site this includes the static and mobile plant as well as portable power tools and hand tools. Examples in the office environment include all electric goods such as kettles, photocopiers, shredders, printers and computers.
- The safe use, handling, storage and transport of all articles and substances.
- To provide such information, training and supervision as is necessary to ensure employees can carry out their tasks safely.
- To ensure that access and egress to and from the work are safe and without risks to health which means keeping these routes, well maintained, clean and clear from obstructions.

- To provide a safe working environment and the provision of adequate welfare facilities.

Employers are also required to prepare and update a written safety policy (section 4.12.2), consult with any recognised safety representatives and to establish a safety committee (section 4.13).

Section 3 – General duties of employers and self-employed to persons other than employees

Employers must ensure that any persons not in their employment who may be affected by the work are not exposed to health and safety risks. This includes official visitors and those delivering goods to the site as well as third parties passing by. Note that there is a duty of care to trespassers. Self-employed persons also have a duty to ensure that other persons are not exposed to risk as a result of their actions. In both cases they have an obligation to provide information about their work where it might affect the health and safety of others.

Section 4 – General duties to persons concerned with premises to persons other than their employees

This is similar to Section 3 except that it is concerned with third parties other than employees, who use the premises for work. Particular attention needs to be given to plant or substances provided and access and egress.

Section 5 – General duties of persons in control of certain premises in relation to harmful emissions into the atmosphere

It is the duty of the owner of any premises, and this includes the contractor during construction, to use the best practicable means for preventing the emission into the atmosphere from the premises of noxious or offensive substances or rendering them harmless and inoffensive before being emitted. At first reading, it appears that this is not a problem for construction sites as it might be thought to be about chemicals, odorous gases and so on, but the interpretation also includes, smoke, grit, dust and noise, of which construction sites are major generators.

Section 6 – General duties of manufacturers etc., as regards articles and substances for use at work

It is the duty of any person who designs, manufactures, imports or supplies any article for use at work to ensure it is safe to use when used properly. This means testing and research has to be carried out to ensure risks to health and safety have been eliminated or reduced to a minimum. Those who install the article must ensure it is safe to use when it is handed over. They must provide adequate information to the user to ensure it can be used safely. There are implications for the main or managing contractor in construction in ensuring this is carried out. Most buildings have equipment installed as part of the building services element, almost invariably installed by sub-contractors who will have sourced the equipment from some other supplier. The systems of control therefore have to be robust.

Section 7 – General duties of employees at work

All employees have an obligation to take reasonable care for the health and safety, not just of themselves, but also for others who may be affected by anything they do or don't do at work. For example, leaving a piece of wood with a protruding nail for others to walk on, spilling coffee on the floor and not wiping it up, ignoring the risk of someone slipping. Employees are also obliged to co-operate with the employer to enable them to comply with their legal obligations.

Section 8 – Duty not to interfere with or misuse things provided pursuant to certain provisions

It is an offence for anybody to deliberately or recklessly interfere with or misuse anything that has been provided in the interests of health, safety or welfare.

Section 9 – Duty not to charge employees for things done or provided pursuant to certain specific requirements

Employers are not permitted to charge an employee for doing anything, or for the provision of safety equipment and clothing, that is required to carry out work safely.

Enforcement

There are three stages of enforcement at the disposal for use by the factories inspector on a standard visit or as a result of a visit resulting from an incident occurring at the workplace. The first is called an improvement notice that details a specific period of time in which the contravention must be corrected. The length of time will be a function of the future risk associated with the hazard and the cost and time it would take to rectify the situation. The second, called a prohibition notice, is where it is judged that there is a serious risk of injury if the activity is continued. In this case the activity will be stopped until the specified remedial action has been completed. Finally, the HSE will prosecute if it is considered the infringement is as a result of negligence. This can result in a fine of up to £5,000 for most offences and £20,000 for more serious offences, although there is no limit to the fine on conviction. The courts have the authority to send the guilty party to jail for up to two years.

4.9.2 Reporting of Injuries, Diseases and Dangerous Occurrences Regulations 1995 (RIDDOR) (adapted from *The RIDDOR Explained*, HSE)

RIDDOR requires that any incident covered by the regulations that occurs on a construction site, has to be reported to the HSE Incident Contact Centre, by phone, fax, the Internet, email or by post depending on the nature of the incident. It is the responsibility of the principal contractor to do this. If it occurs in the head or regional office in the UK, then it can be reported to the environmental health department of the local authority. This reporting procedure gives the HSE the opportunity to identify where risks arise and to investigate serious accidents as well as provide statistical evidence on the safety performance of all types of commerce and industry and this in turn informs the legislature in making decisions about new or revised regulations.

A report is required in the following circumstances:

Death or serious injury

If a directly or indirectly employed person, such as by a sub-contractor, is killed or suffers a major injury (including any caused by physical violence), or a member of the general public is killed or taken to hospital then the enforcing authority must be notified without delay, usually by telephone.

This must be followed up within 10 days by a completed accident report form (F2508). Examples of a major injury are specified in the regulations and include:

- fractures other than to fingers, thumbs or toes
- amputation
- dislocation to shoulder, hip, knee or spine
- loss of sight
- any serious injury to the eye
- unconsciousness caused by asphyxia or exposure to harmful substances
- acute illness requiring medical treatment resulting from inhalation, ingestion, absorption through the skin and exposure to a biological agent or its toxins.

Over-three-day injury

An over-three-day injury is classed as one that is not major, but results in the injured person being away from work or not capable of carrying out the full range of normal work activities for more than three days. This does not include the day of the accident, but does include weekends, rest days and holidays. In this case the incident has to be reported on the same form as above (F2508) within 10 days.

Disease

If a doctor notifies the employer that a worker has a reportable work-related disease, a completed disease report form (F2508A) must be sent to the enforcing agency. The full list can be obtained from the HSE, but examples include:

- certain poisonings
- skin diseases, such as dermatitis and skin cancer
- lung diseases such as asbestos-related
- infections such as tetanus
- other conditions, such as certain musculoskeletal disorders and hand-arm vibration syndrome.

Dangerous occurrences

When an incident occurs not resulting in a reportable injury, but which could have done, this could be classed as a dangerous occurrence and has to be reported immediately and followed up within 10 days with form F2508. Examples of dangerous occurrences are:

- a collapse of part of the building during its construction, alteration or demolition; this would include the temporary works such as scaffolding and formwork
- major collapse of the soil into an excavation
- plant coming into contact with overhead power lines
- malfunction of breathing apparatus
- a sudden uncontrolled release of flammable liquids and gases above a level set by the HSE
- accidental release of any substance which may damage health.

Records of any of the above reportable incidents have to be kept by the employer for three years after the date on which it occurred. This includes the date and method of reporting; the date, time and place of the incident; the personal details of those involved; and a brief description of the nature of the incident or disease.

4.9.3 The Control of Substances Hazardous to Health Regulations 2002 (COSHH)

At first glance it may be thought the construction industry does not use many materials or chemicals that are hazardous to health, but on closer inspection not only are many chemicals such as adhesives, timber preservatives and solvents common, but seemingly innocuous materials such as some species of timber and concrete can, whilst being processed, cause dusts which are hazardous. Further, processes such as welding give off hazardous fumes, and waste at the end of the process can create a hazardous situation on disposal, such as the residue in containers. There are exemptions to the regulations such as asbestos, lead and radioactive materials as these have their own regulations.

There are eight steps required under the regulations, but often in practice they are not all carried out as diligently as they should. The first step is to assess the risk, which means identifying all hazardous substances present in the workplace, and, in the case of construction, all those that are to be purchased to complete the project. The hazards associated with many of the materials are identified with a warning label on the containers and all suppliers must provide a safety data sheet for them. Assessing the risks is about using experience and common sense in establishing how likely it is that the material can affect the health of the operatives and third parties, when used for a particular application in a specific location and over what period and how frequently. For example, where fumes are the potential hazard, is the place of application well ventilated or enclosed?

Having established the type and nature of the hazards and ranked their risk, the second step is to decide what precautions need to be taken. If significant risks are identified, a plan on how to remove or reduce the risk needs to be implemented. A risk is seen to be significant if the published workplace exposure limits (WELs) are lower than that which the worker is exposed to or the method of work does not comply with usual construction industry good working practices and standards. A WEL is the maximum concentration of an airborne substance to which workers may be exposed to by inhalation. It is necessary to take note of the supplier's advice on correct use, handling, storage and disposal recommendations. Besides inhalation, substances can also enter the body orally and through the skin and this risk should also be considered. Finally, the company's control and monitoring systems must be working and are shown to be effective and this includes recording and maintaining a record of the decisions made.

If there is little or no risk, then no further action need be taken, but if there is, then step three and onwards comes into effect; to prevent or adequately control exposure to hazardous substances if it is reasonably practical to do so. A simple rule is that if you have two substances that perform equally well and are of similar in cost, then use the less toxic. Far too often, the safety data sheets and warnings on the containers are accepted and only the precautions necessary to satisfy these needs are taken. Considering alternative methods of executing the process can also result in not using the substance, or the materials could be used in an alternative form or applied differently such as with a paintbrush rather than a spray gun. If prevention is not possible or reasonably practicable, then the exposure levels have to be controlled. This can be achieved by totally enclosing the process, which is difficult in construction activities as they almost invariably require hands-on labour; partially enclosing the process and using mechanical ventilation extraction equipment; ensuring there is good natural ventilation; reducing the numbers of workers exposed to the substance, especially those who are not needed in the process such as those employed on an other trade; and developing safer methods of work so that the likelihood of the materials leaking or spilling is minimised. The last resort is to use personal protective clothing if the aforementioned methods are insufficient. These include respirators, face masks and other protective clothing such as gloves, aprons and full body coverings. The reasons for all these different approaches is the same in all cases: to reduce the exposure to a level that a normal healthy worker can safely work at every day without their health being adversely affected. The levels of acceptable WELs vary depending on the material and are found on the safety data sheets.

It is a requirement under the regulations that employees use the control measures installed and report if they are not working properly, and management has a responsibility to see that they do and are encouraged to do. This is step four, and requires that all equipment, personal protective clothing and procedures should be checked at appropriate intervals. Whilst not common in the construction work environment, step five requires that if the risk assessment concludes there could be a serious risk to the health of the employees if the control measures put in place fail or the WELs could be exceeded, then it is necessary to monitor the exposure levels and keep those records for a minimum of five years. In step six, if there is a risk that a substance which the employee has had contact with is linked to a particular disease or adverse health effect, it is necessary to carry out health surveillance. These records have to be kept for 40 years.

In step seven, if there is a risk of an accident that is higher than the risks associated with normal day-to-day work, the employer must have a plan in place to respond to such an event. This is a sensible approach to all potential emergencies, even when not covered by the COSHH regulations. There should be safety drills carried out from time to time to ensure that in the event, those involved are properly prepared. This is especially important in construction as the workforce is continually changing.

Finally, in step eight employees involved in the use of hazardous materials and processes must be properly supervised, trained and informed about the materials they are using. Employees cannot be expected to carry out their role of using the control measures (step four) if they are not properly trained and informed. They need to know what the hazard is, how risky the employer believes it to be, what work procedures have to be complied with, including the use of protective clothing and equipment, and in the event of an accident, what the emergency procedures are.

4.9.4 Construction (Design and Management) Regulations 2007

These regulations, commonly referred to as the CDM regulations, require that health and safety issues be considered at every stage of the process including design, construction, alterations, repair and maintenance. This means that everybody involved in the process including the client, designers and the managers of the construction processes, is covered by the regulations.

In the original 1994 regulations, two new roles were created, appointed by the client, called the planning supervisor, responsible for the co-ordination of the health and safety aspects of the design and planning phase, and the principal contractor, responsible for the planning, control and management of the health and safety of the construction process. However, the role of the planning supervisor has been replaced the CDM co-ordinator in the 2007 regulations. In summary, the regulations identify responsibility for the different people listed below:

The client must appoint a competent CDM co-ordinator, as soon as is practical, after initial design work or other preparation for construction work has begun. They must also appoint a principal contractor for the duration of the contract and commissioning stages, and provide adequate resources for health and safety. They must also ensure that all designers and contractors

Table 4.3 Pre-tender plan

Description of the project
Particulars, including a description of the project; details of the client, design team and consultants, CDM co-ordinator; the extent and location of existing records and plans
Client's consideration and management's requirements
The organisation, its safety goals and arrangements for monitoring; permits and authorisations requirements; emergency procedures; site rules and other restrictions on the contractors and suppliers; activities on or adjacent to the site during the works; methods of communications; security arrangements
Environmental restrictions and existing on site risks
a) *Safety hazards* Boundaries and access, including temporary; adjacent land uses; existing storage of hazardous materials; location of existing services; ground conditions; existing structures and their current state b) *Health hazards* Asbestos, including results from surveys; existing storage of hazardous materials; contaminated land, including surveys; existing structures hazardous materials; health risks arising from client's activities
Significant design and construction hazards
Design assumptions and control measures; arrangement for co-ordination of ongoing design work and handling design change; information on significant health and safety risks identified during the design; materials requiring particular precautions
Health and safety file
See section 4.11

employed on the contract are competent and provide them with adequate resources to take account of health and safety issues. In all cases they must ensure that all involved have sufficient time to carry out their work properly. The contractor must not be allowed to commence work until the principal contractor has prepared a satisfactory health and safety plan and ensured that the health and safety plan is available for inspection after the project is completed (see Table 4.4).

The designer must ensure that the client is aware of their responsibilities under the regulations since not all clients are knowledgeable about the construction industry. They should critically assess their designs throughout the design process, including any modifications made later, with a view to identifying any hazards in the proposed design and trying to eliminate or reduce accordingly (section 4.9.3). This includes risks likely during the construction

Table 4.4 Contract phase plan

Description of the project
Particulars, including a description of the project; details of the client, design team and consultants, CDM co-ordinator; the extent and location of existing records and plans
Communication and management of the work
Management structure and responsibilities; health and safety goals and arrangements for monitoring; arrangements for regular liaison between parties on site; consultation with the workforce and the exchange of design information between the client, designers, CDM co-ordinator and contractors; handling design changes; selection and control of contractors; exchange of health and safety information between contractors; security, site induction and on-site training; welfare facilities and first aid; reporting and investigation of accidents and incidents including near misses; production and approval of risk assessments and method statements; site rules; fire and emergency procedures
Arrangements for controlling significant site risks
a) *Safety risks* Services, including temporary installations; preventing falls; work with fragile materials; control of lifting operations; dealing with services, e.g. water, gas; the maintenance of plant and equipment; poor ground conditions; traffic routes and segregation of vehicles and pedestrians; storage of hazardous materials; dealing with existing unstable structures; accommodating adjacent land use; other significant safety risks b) *Health hazards* Removal of asbestos; dealing with contaminated land; manual handling; use of hazardous substances; reducing noise and vibration; other significant health risks
Health and safety file
Layout and format; arrangements for the collection and gathering of information; storage of information

work as well as those involved in maintenance of and alterations to the completed building, and demolition. Where the hazard cannot be completely designed out, this has to be stated to the planning supervisor who can include it in the health and safety plan (Table 4.4). Examples of designing out hazards include specifying less hazardous materials such as solvent-free or low-solvent materials; avoiding processes that generate fumes, dust and noise; specifying materials that are easier to handle; consideration of off-site manufacture; and solutions that allow the operatives to work from inside the building rather than from scaffolding and ladders, such as when cleaning windows.

The CDM Co-ordinator ensures that all those involved in the design work collaborate on safety matters and eliminate or reduce the hazards in their design. They must ensure that the HSE is notified of the project. They are responsible for producing the pre-tender health and safety plan (Table 4.3) and they should highlight or assist the designers in identifying risk and subsequent reduction, including any subsequent design changes. They should be able to give advice to the client on the competence of the designers and principal contractor in matters pertaining to health and safety, but not in their design capability. They should also advise the client on the adequacy of the construction phase health and safety plan (Table 4.4). At the end of the contract they must ensure that the health and safety file is prepared and handed to the client.

The principal contractor's key responsibility is to manage and co-ordinate the construction phase health and safety issues. They are usually the main or managing contractor and they are responsible for vetting the competence of any designer or contractor they employ in terms of health and safety compliance. They must ensure that a construction phase health and safety plan is prepared before construction begins, that it is implemented, reviewed and actual performance monitored against it (Table 4.4). They have to promote co-operation between all the contractors employed on site, see that unauthorised people do not enter the site, and enforce site rules. They have to see that people on the site receive appropriate training and information about safety, and the workforce is consulted on safety and health issues.

They must clearly display the notification of the project to the HSE and ensure all the contractors are made aware of its contents. The notification is a form which gives the names and addresses and telephone numbers of the client, CDM co-ordinator and principal contractor, the address of the site, the local authority area, approximate start date, duration of the contract, the maximum numbers of workers at any one time and contractors to be employed, a brief description of the works, and the name of the main contractor. This is then signed by the CDM co-ordinator and the principal contactor.

Contractors must not begin work on site until they have been provided with the names of the CDM co-ordinator and the principal contractor, and the parts of the health and safety plan relevant to them. If they find errors or shortcomings in this part of the plan they should notify the principal contractor. Once on site they must satisfy themselves that any designers or contractors they employ are competent and provide information to the principal contractor about risks to others created by their work. They must comply with the rules of the site and advise the principal contractor of any accidents or dangerous occurrences that occur.

4.10 Health and safety plans

The approved code of practice in *Managing Health and Safety in Construction* (HSE 2007) states 'the health and safety plan should include or address all the following topics where they are relevant to the work proposed'.

It is easy to forget that health and safety issues should be thought about during the design process and it is not just a construction process problem as this is where the accidents occur. However, thought and careful planning during the pre-tender phase can assist in the overall reduction of accidents and is in line with the philosophy underpinning the CDM regulations. Its purpose is to provide information for those bidding for the work, which can include both the main contractor and sub-contractors. The construction phase plan is to show how safety and health will be managed during this process. There are some similarities between the two as is demonstrated in Tables 4.3 and 4.4, adapted from *The Managing Health and Safety in Construction* leaflet (HSE 2007).

When drawing up both of these plans it saves time later if the CDM co-ordinator and principal contractor consider the issues that will have to be used in the health and safety file which has to be handed over to the client or the building's end-user.

4.11 Health and safety file

It is the responsibility of the CDM co-ordinator to produce this file which is a record of information for the client or the end-user, alerting them to any health and safety matters they might have to manage in the future during subsequent maintenance, alterations and extensions. The file includes:

- A brief description of the work carried out including drawing and plans of the building and works.

- Any information about residual hazards and how they have been dealt with, such as any surveys in refurbished buildings showing asbestos which has not been removed or has been repaired.
- Data on structural design such as safe working loads for floors and roofs and identifying elements that have been pre- or post-tensioned.
- Hazards associated with any materials that have been used which could cause a problem if not maintained or removed in a certain manner.
- Information advising on the methods needed to remove or dismantle installed plant and equipment.
- Manuals outlining the safe operating and maintenance procedures for installed plant and equipment.
- Details of the location and types of services, including emergency and fire-fighting systems.

4.12 Managing safety in the construction industry

4.12.1 Introduction

Whilst everybody in an organisation has a responsibility for their safety and that of their colleagues, it is essential management takes ownership of safety at the highest level, otherwise it is unlikely that the best safe working environment will be achieved. The more senior the management, the more likely they are in ensuring standards are set, maintained and improved. The HSE would like one director to be made responsible for safety so that companies cannot hide behind 'corporate responsibility', which makes it more difficult to obtain a conviction. However, there has been opposition to this. Some companies have been considering making the annual bonus of their board members determined in part, if not in whole, by their section's or the business's safety record.

The key elements of successful health and safety management are shown in Figure 4.2 are taken from *HSG 65: Successful Health and Safety Management* (HSE 1997).

4.12.2. Safety policy

The first stage in any safety strategy is to produce a safety policy. Its prime purpose is to improve the safety of the work environment by reducing the number of injuries and accidents and engaging the workforce in becoming safety conscious. It is a published statement reflecting the organisation's intentions in relationship to the management of health and safety matters. It

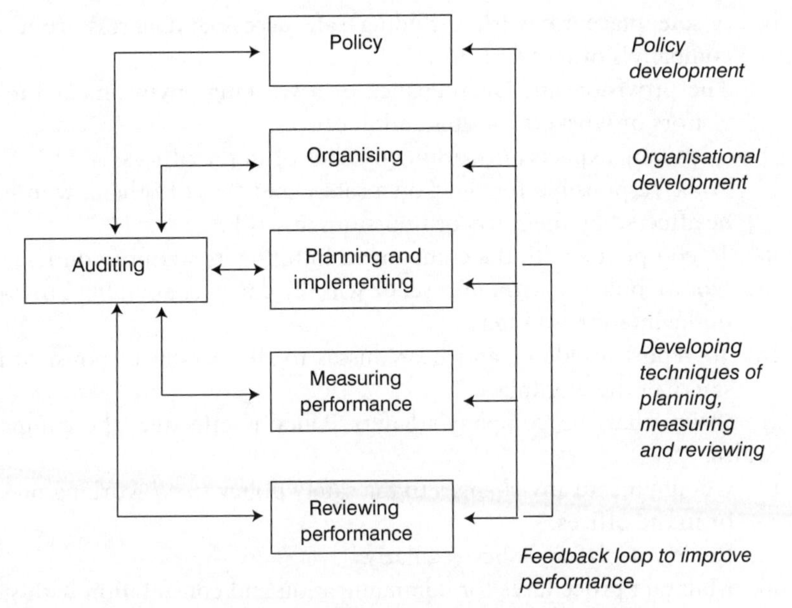

Figure 4.2 Key elements of successful health and safety management

should also define the organisation's corporate philosophy towards health and safety matters in the context of the business activities and be clearly presented in the form of a policy statement, originating from the board of directors. It is against this that the performance of the organisation can be compared. A typical policy would include similar statements to the following:

1. The company is committed to providing a safe working place for its entire staff. The company will pursue a policy of continual improvement in its safety management and will review legislation and introduce procedures as required to meet its safety requirements.
2. It is the company's intention to take all steps necessary as are reasonably practicable to meet its responsibilities in providing a safe and healthy environment for its entire staff, visitors, suppliers, contractors and the general public.
3. The company will seek to provide:
   i. Systems of work that are safe and without risk to health.
   ii. Safe arrangements for the use, handling, storage, transport and disposal of articles and substances.
   iii. Adequate supervision, instruction and training to enable all employees, contractors and visitors to avoid hazards and to contribute to their own safety and that of others.

- iv. A safe place to work including safe access and egress from the company's offices and sites.
- v. The provision and maintenance of a working environment for *all* visitors or workers on sites and in offices.

4. The company expects *all* working on sites or in the offices:
   - i. To be responsible for their own safety and that of others, who may be affected by their acts or omissions at work.
   - ii. To co-operate with the company in fulfilling its statutory duties.
   - iii. Not to interfere with, misuse, or wilfully damage, anything provided in the interests of safety
   - iv. To report accidents and near misses to the person responsible for safety in the workplace.
5. To ensure that the company's Safety Policy is effective, the company will:
   - i. Communicate any changes to the safety policy to *all* working on site or in the offices.
   - ii. Review its Safety Policy regularly.
   - iii. Maintain procedures for communication and consultation with staff on matters pertaining to safety.

Often accompanying the published statement will be further information outlining the company safety organisation structures and the responsibilities of those involved in safety matters, the arrangements for identifying hazards, assessing risks and controlling them.

This document needs to be distributed within the organisation to all current and new employees, but if this is all that is done, it will probably just be filed away and forgotten by the majority of the recipients. It is important all employees attend some form of training session so not only is the document introduced to them, but their part in ensuring a safe and healthy working environment is made clear. The document must always be kept up to date and staff notified of any changes.

4.12.3 Organisation of safety and lines of communication

The important issue is to demonstrate to the employees of the company how seriously it takes safety. This involves the active involvement of the board and having lines of communication so senior management can react rapidly. Figure 4.3 demonstrates a typical mechanism for this to happen. In this case, the managing director is made responsible for co-ordinating safety issues, but it could equally be another member of the board.

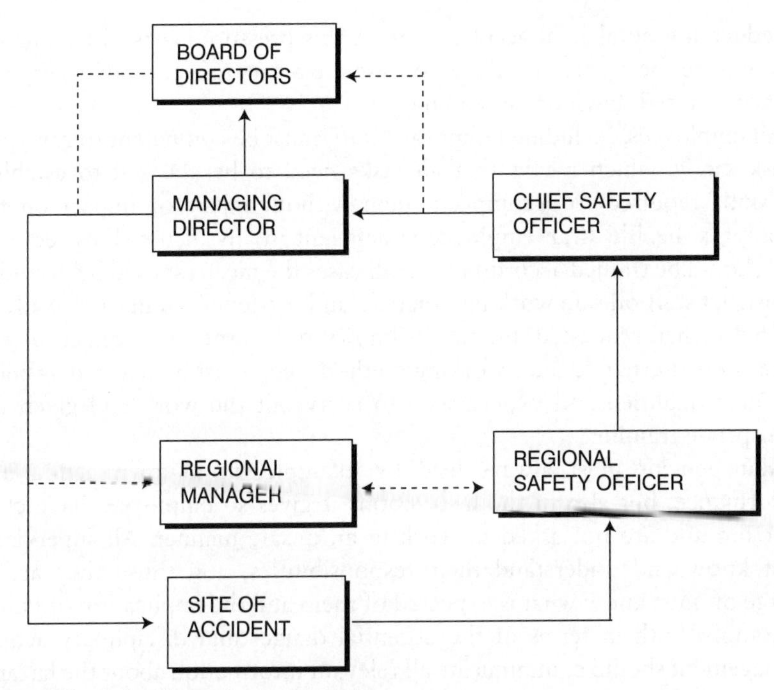

Figure 4.3 Safety lines of communication

If an accident occurs on site then the procedure is followed depending on the severity of the accident. Minor reportable accidents are reported to the regional safety officer, usually on a monthly basis, who then collates all the information from all the sites in the regional and submits to the chief safety officer who presents the whole company record to the managing director or board.

Over-three-day injuries (section 4.9.2) are immediately reported to the regional safety officer at the end of the three days or at the end of month and then to the board via the chief safety officer. Death, serious injuries and dangerous occurrences, as defined for RIDDOR, are notified immediately by phone to the regional safety officer who quickly notifies the regional director and the chief safety officer who, in turn, immediately advises the managing director. All reports of accidents will also be sent to the regional director on the same time scale.

Under most circumstances, in a large organisation, the site management rarely has direct contact with the senior members of the company other than site visits, but in this case senior management will contact the site directly if the safety performance is unsatisfactory or if it is excellent. Site management naturally wishes to avoid the former and therefore becomes focused on trying to reduce the number of accidents. Since this pressure exists, the company must ensure the records produced on site are a true reflection of reality and are not watered down or misleading.

All employees, including temporary staff, must be competent to carry out a task safely, which means that all tasks need to be assessed to establish the skills required to accomplish them without injury or impact on the employees' health. If an employee is deficient in any of the skills required they should be trained accordingly. In all cases the means should be provided to instruct staff on safe working practices and procedures and to give advice and help when requested. If a task is known to be especially dangerous and there is no alternative safer working method, they must be selected as being the most qualified and experienced to carry out the work and given the appropriate training.

Management must always lead by example, in its own action and performance, but also in the instructions it gives so employees have clear direction and are not asked to work in an unsafe manner. All supervisors must know and understand their responsibilities, and those they are in charge of must know what is expected of them and the implication of failing to comply, both in terms of the potential danger and disciplinary action. Management should communicate all relevant information about the hazards and risks associated with the task and identify the precautionary measures to be taken.

The safety of any work situation can be improved if the staff are involved and consulted as often they have a more detailed understanding of the task in hand. They should be involved in planning the way the work is to be carried out and in resolving problems at the planning stage and when the task is being executed. If the task has to be repeated and requires written procedures, they could contribute to this process as well. Health and safety issues should be discussed regularly either in a formal setting, if necessary, or as an integral part of the day-to-day management of the works.

4.12.4 Planning, implementing and setting safety standards

The first stage in the plan is to identify the hazards and assess the risks. On a construction project this should be carried out at the tendering stage and when the contract is awarded. The production of the method statement is an ideal place to start as the production process is being looked at in detail for the first time and where level of risks involved could have an impact on the price. If the risk is high for any part of the process, it may mean a change of method so as to eliminate or reduce the amount of risk. To this is added information given by the CDM co-ordinator generated from the design process. Once the contract is awarded, further detailed hazard identification can be made from the latest information available. Risk assessments are covered in more detail in section 4.14 and *Business Organisation for Construction*, Chapter 10.

Having established the risk likely from the hazard and confirmed there is no alternative method, a detailed analysis of the process and the safety precautions that have to be invoked, takes place. Any method developed must comply with the current health and safety laws and regulations that are appropriate to the business on site and in the offices. This involves protective clothing and equipment needed, limitations on how the operative can work, such as the positioning of guards on cutting equipment, and the protection of third parties such as when welding. From this can be produced standards of performance against which monitoring can take place (section 4.12.5). These health and safety targets should be discussed with the managers, supervisors and all the sub-contractors and agreed that they are practical and achievable. On occasions, the work on the construction site can impact on the safety and health of neighbours and, in the case of alterations and extensions to existing premises, to those working in close proximity to the construction process. Where this is identified, third parties should be consulted so that appropriate solutions can be found and implemented. The site requirements should be communicated to the purchasing department so when ordering materials the methods of delivery, packaging and unloading comply with the sites safety requirements.

Whilst the reason for putting all these systems in place is to eliminate accidents, they may still occur, so procedures on how to deal with anything that may go wrong have to be considered. All the procedures must be written down so nothing is left to chance.

Much of this is purely administrative and when implemented will have some effect for the good, but to derive full benefit, the overall culture of the business has to be directed to making safety a normal management instinctive practice and not simply seen as an added task that has to be considered from time to time.

Setting standards is one of the ways that the culture of the organisation can be influenced because it becomes engrained in the minds of the employees. However, this will not happen unless the standards set are measurable, achievable and realistic. Wherever there are standards commonly in use it would be usual to adopt them (e.g. HSE regulations), but where not these have to be developed, taking advice where appropriate.

Standards should be set and applied across the full range of the business's activities. The workplace includes the site, offices, workshops and the regional and/or head offices. Standards are not just about the methodology of the work, but involve environmental controls such as ranges of temperature, humidity, lighting and dust levels. It includes ergonomic issues such as seating and proximity and ease of materials to hand (section 3.2.4). There are special issues regarding the use of plant and equipment which often require designated training such as when using percussion hand tools or walking along the jib of a tower crane. Mobile plant can potentially collide with people or other objects and must be stored so nobody can operate it other than the designated experienced operative. Methods and the frequency for checking guards on machinery or scaffolding guardrails and toe boards, etc., have to be considered. Raw and unfinished materials have to be off-loaded, stored, transported to the place of work, processed and unused or scrap material recycled or tipped. If the process produces hazardous waste or toxic emissions, the maximum acceptable levels have to be set, as have levels of emissions of dust and noise. The hazard and risk has to be assessed for each of these operations. Finished components have to be unloaded, either stored before being placed in the building or lifted directly into position requiring specialist lifting gear and access. The systems for assessing design solutions, especially amendments during construction, need to be in place. Standards and targets need to be set for the training of all those employed in the workplace and, where relevant, visitors. Agreements should be in place to consult staff, or their representatives, at specified intervals.

It is easy to fall into the trap of having generalities in the descriptions of the standards and targets, which are difficult to either, measure or achieve. For example:

- 'Staff must be trained' is a meaningless statement as it does not define what training means and involves, nor who carries it out.
- 'All machines will be guarded', on the surface appears to be specific, but how is this to be achieved and to what standard is the protection, as the ultimate protection will probably make the machine unusable.
- 'All employees will wear safety helmets' again seems reasonable, but does this mean they have to wear them when sitting at their office desk and is it acceptable if the helmets are ten years old?

4.12.5 Measuring performance

As in any other part of the business it is necessary to measure performance to discover whether or not the company is successful. The method of monitoring has to be determined, as does the frequency at which it takes place. The frequency is a function of the level of risk and the temporary nature of its use. For example, scaffolding can often be changed to suit production requirements and because the risk of people falling from heights or dropping objects is high, the frequency of monitoring would be much higher than, say, checking to see whether or not the kettle in the office is still electrically sound.

The aim of monitoring is not just to ensure that targets are being met, but also to continually improve performance. There are three basic sets of questions that need to be addressed when monitoring performance. They are:

1. *Where* are we now? In other words are we reaching the target set?
2. *Where* do we want to be? The answer to this is either on or approaching target, or if having attained targets, to raise the threshold further, but to achievable and realistic levels.
3. *What* is the difference and *why?* If below target it is essential to establish the reasons for this.

There are two types of monitoring of performance: active and reactive. Active monitoring is monitoring before things go wrong. This is seeing if the standards set are being implemented and then monitoring their effectiveness. Reactive monitoring occurs after things have gone wrong. Examples of this include investigating the causes of injuries and dangerous occurrences, cases of illness, when property has been damaged and near misses. In each case, identifying why the performance was sub-standard and then instigating change, if necessary, to improve the situation for the future.

4.12.6 Learning from experience: auditing and reviewing

Whereas monitoring provides the information to review activities and decide how to improve performance, auditing complements monitoring by looking to see if the policy, organisations and systems are achieving the correct results. A combination of the results from measuring performance and auditing can improve health and safety management.

The key issues to look for are whether the results demonstrate a compliance with the standards set; if there are standards to measure against; if the standards are set too low; and if the improvements set have been achieved in the time scale set. Finally, an analysis of the injury, illness and incident data can identify any underlying trends, their causes and if there are any common features such as a high occurrence of back injuries.

It is important that the company learns from the mistakes rather than giving excuses for the causes. It should take action as a result of the audit findings and this information be fed back to advise future policy and improve performance by increasing the standards.

4.13 Safety committees

The Safety Representatives and Safety Committee Regulations 1977 (SRSCR) came into effect in October 1978, and were modified by the Management of Health and Safety at Work Regulations 1992. These regulations and Section 2(7) of the HSW, provide for the appointment of safety representatives by recognised trade unions, and the setting up of safety committees at the request of these representatives. There is no reason why the company cannot set up safety committees where there is no trade union representation, nor do they have to wait for a request from the safety representatives. Under the SRSCR, the trade union safety representatives are to investigate possible dangers at work, the causes of accidents and general complaints by employees on health and safety and welfare issues, and to take these matters up with the employer. They should carry out inspections of the workplace particularly following accidents, disease or other events. Besides attending safety committees, they represent employees in discussions with health and safety inspectors. In carrying out this role they are protected against dismissal or other disciplinary action when taking part in health and safety consultation.

The composition of a safety committee depends on the way it is set up. To work effectively there needs to be representation from the majority, if not all, of the facets of the workplace. Clearly, in a very large industrial works, the committee may become too large and unwieldy if every part of the business is represented, although there is no reason why there cannot be safety subcommittees reporting to the main committee, to deal with this problem. If the employees are trade union members, then the membership of the committee will be selected through the trade union appointment process, but if not, then non-trade union members must be consulted by their employers. On construction sites with high percentages of sub-contractors, representatives from the key trades should be enlisted.

The guidance to the SRSCR states that the size, shape and terms of reference of the committee must depend on discussion and agreement between employers and unions. The recommendations are that the committee should be compact; there should be a 50/50 split between management and employee representatives; safety officers and any other advisors should be ex-officio members; and that safety committees could also provide a link with the enforcing authorities such as the HSE and local environmental health officer. There is nothing to stop the split between management and employee representatives being different to the 50/50, but it is suggested this should only be in favour of the employees.

The success of a safety committee depends largely on the attitude of management. If they go in with a positive attitude and demonstrate their willingness to listen, accept reasonable arguments for improving safety, and then enact the changes, the committee can make a successful contribution to safety performance improvement. It is not unusual for a member of the management team to chair the meetings because he or she will have chairmanship skills and experience, but this is not mandatory.

A typical agenda for a safety committee could include the following items:

- Reviewing the progress of recommendations made at the previous meeting(s). This acts as demonstration of management's willingness to listen and act.
- To discuss the causes of accidents or incidents that have arisen since the last meeting.
- To inspect and review accident and ill-health trends with a view to improve working methods and reduce the number of accidents. This highlights the overall trends, but also can be used as a way of establishing if there are similar accidents that regularly occur which can focus the members on resolving this issue. Illness and absenteeism, if confined to a particular work activity, can indicate higher than normal levels of stress, but may also be about the quality of the management. Sometimes an individual's name comes up with regularity and if not perceived as a malingerer, needs special attention as the operative may require special training.
- Examine the health and safety implications of the next phase of construction, or of a new piece of plant or equipment that is going to be introduced.
- Examine the safety audits and make recommendations.
- Discuss the reports made by the safety representatives.
- Review the content of the health and safety training for employees, subcontractors and visitors.
- Review risk assessments, especially if a new phase of the construction work is about to commence.
- Consider any reports and information of relevance sent by the HSE.
- Review any safety publicity and campaigns and their effectiveness.

The frequency of the meetings depends on the nature, size and rate of changing activities of the work, but once a month would not be unreasonable.

4.14 Instruction and training

Instruction means telling people what they should and should not do, whereas training means helping them learn how to do it, with the emphasis on helping and learning.

The first stage is to establish who needs to be trained and this starts with the person initiating the training programme. Have they the full understanding of the implications and needs of the organisation in terms of the legal requirements, its current culture and image, what the company wishes to achieve and why, and how this can be brought about? The managers and supervisors have responsibility for the safety at the workplace, designers need to understand their responsibilities because of the CDM regulations, new recruits have special needs as do part-time employees. In the office, estimators and planners have to comprehend the safety implications of any methods and sequence of operations they adopt, and purchasing officers have to understand how materials are handled and stored on site safely. Should the company also be involved in training sub-contractors, especially if their organisation is small and does not have the knowledge or resources to carry it out themselves?

The next stage is to determine what training is needed and the outcomes expected in terms of the knowledge and experience needed to work safely when carrying out their role, i.e. levels of competence. To give people the wrong training, or too much, is a waste of time and money. This has to be designed to take account of the current state of the company's safety performance and should consider the need for refresher training and updating of existing skills and knowledge, especially if the regulations change or recommendations are made by the enforcing authorities.

Having decided on the need, the method of training has to be determined. There are two approaches, first to use in-house trainers, providing they are competent in health and safety matters and second, to use the expertise of external training providers such as the Construction Industry Training Board (CITB).

It is necessary to prioritise training needs and produce a plan of training. This has to be costed to ensure that adequate resources are made available. The CITB may contribute towards training costs. There are key times and occasions when employees should be trained. These are when they start working for the organisation, if they have a work or responsibility change, if they have not used their skills for a while, and if new or changed risks occur at the workplace.

To complete the cycle, the effectiveness of the training must be monitored to establish if the standards of competence set have been achieved and if there

Table 4.5 Typical hazards on a construction site

Working at heights	Electrocution
Falling from heights	Mobile plant and machinery
Objects falling from heights	Hand tools
Excavation collapses	Volatile organic compounds (VOCs)
Asbestos	Nails poking through wood
Temporary works	Chemicals, adhesives and wood preservatives
Operatives not used to working together	Dust, fumes and fibres
Fragile roofs	Ladders
Handling heavy materials	Loose reinforcement tie wire

has been any improvement in the company's health and safety performance. A system of permitting feed back from managers and supervisors and those who have been trained, contributes to improving future performance.

4.15 Risk assessment

Risk assessment is discussed in fuller detail in *Business Organisation for Construction*, Chapter 10, but as it is an integral part of the improving safety in construction, it would be unfortunate not to include it here. There are three key elements in assessing risk in safety: hazard identification, evaluating the risk, and installing preventative and protective measures.

4.15.1 Hazard identification

As indicated earlier, hazard is something that presents a potential to cause harm and construction, is by its very nature, a process where there are a high number of hazards. Table 4.5 gives an indication of just a few of the many typical hazards to be found on a construction site.

It is necessary to systematically assess all the production stages of the process and assess where potential hazards are. This also means calling in the safety data sheets required for COSHH (section 4.9.3).

4.15.2 Evaluation of risk

Once the hazards have been identified, the risk is assessed using one of the various methods available. There are two factors that affect the degree of

Table 4.6 Criteria for assessing risk

Criteria for measuring hazard severity		*Criteria for likelihood of occurrence*	
Description	*Assigned value*	*Description*	*Assigned value*
Minor injury requiring no first aid attention	1	Remote: highly improbable	1
Accident needing first aid attention	2	Unlikely: in exceptional circumstances only	2
Over-three-day reportable injury	3	Possible: where certain circumstances might influence occurrence	3
Serious injury	4	Likely	4
Death	5	Probable	5
		Highly probable	6

risk: the level of harm that might occur as a result of the hazard, and the likelihood of the occurrence or the frequency at which it might occur. For example, crossing a busy road is clearly a hazard and the amount of harm received as a result of being hit by a car is high, but the likelihood of this occurring if using a pedestrian crossing, reduces the risk considerably. If the severity of the hazard is multiplied by the likelihood of it happening, a degree of risk is produced. To do this criteria have to be chosen for them both. Table 4.6 shows a typical example.

Using the assigned values in Table 4.6, the degree of risk associated with an event that is *likely to occur* resulting in an accident *needing first aid attention* would be: $4 \times 2 = 8$. Whereas an *unlikely* event, but if occurring would result in *death* would be: $2 \times 5 = 10$.

Using these criteria, it can be seen that the maximum degree of risk is 30 (5×6), so the values calculated above of 6 and 10 could be expressed as a percentage of the maximum risk, i.e. 20 per cent and 33.33 per cent, respectively. This calculation is carried out because in risk assessment, a priority rating often is preferred rather than degree of risk. For example:

Low priority (L)	<10%
Medium priority (M)	10–50%
High priority (H)	>50%

Note that the percentages chosen for each category are arbitrary and could be any figures deemed appropriate.

An alternative way is to compare the likelihood of occurrence with the severity of harm the hazard would be if the event took place as shown in Table 4.7. In this case a scale of 1 to 10 is used to categorise the likelihood and severity, 1 being the lowest and 10 the highest. The two numbers are

Table 4.7 Priority of risk

Likelihood of occurrence	*Severity of harm*	*Percentage risk*
Low	High	10
Medium	High	50
High	High	100
Low	Medium	5
Medium	Medium	25
High	Medium	50
Low	Low	1
Medium	Low	5
High	Low	10

then multiplied together to give a percentage risk. The table could be broken down into smaller bands than low, high and medium.

These are but a few ways of assessing risk, but the important issues is it is essential to have some mechanism to do this so that one can focus on potential danger areas in the design and construction processes, during the building's lifetime and final demolition.

References

Bird, F.E. (1974) *Management Guide to Loss Control*. Institute Press.

Coble, R.J., Hinze, J. and Haupt, T.C. (2000) *Construction Safety and Health Management*. Prentice-Hall.

Health and Safety Commission (2004) *Comprehensive Statistics in support of the Revitalising Health and Safety Programmes: Construction*. HSC.

Health and Safety Executive (1997) *Successful Health and Safety Management*. HSG65. HSE.

Health and Safety Executive (1999) *RIDDOR Explained: Reporting of Injuries, Diseases and Dangerous Occurrences Regulations*. HSE.

Health and Safety Executive (2005) *Workplace Exposure Limits: Containing the List of Workplace Exposure Limits for Use with the Control of Substances Hazardous to Health Regulations 2002 (as amended)*. EH40/2005. HSE.

Health and Safety Executive (2007) *Managing Health and Safety in Construction. Construction (Design and Management) Regulations 2007. (CDM) Approved Code of Practice*. L144. HSE.

The Control of Substances Hazardous to Health (Amendment) Regulations 2002

The Health and Safety (Display Screen Equipment) Regulations 1992

The Management of Health and Safety at Work Regulations 1992

The Manual Handling Operations Regulations 1992

The Personal Protective Equipment at Work Regulations 1992

The Provision and Use of Work Equipment Regulations 1992

The Reporting of Injuries, Diseases and Dangerous Occurrences Regulations 1995

The Workplace (Health, Safety and Welfare) Regulations 1992

CHAPTER

Waste management

5.1 Introduction

There was a time when the standard joke of an occupier of a new house was that 'there are more bricks left in my garden than was needed to build my house'. Whilst this was an exaggeration, the reality was that a considerable amount of building materials were wasted. Indeed, the Cambridge University Centre for Sustainable Development has suggested that there is the equivalent of enough waste left from thirteen houses to construct a fourteenth. This is no longer acceptable and the industry increasingly has woken up to the fact that it is expensive to create waste and environmentally unacceptable.

The UK construction industry consumes the equivalent of six tonnes per person each year. In the USA it is nearer ten tonnes. It is estimated that the total annual waste created in the construction and demolition stages in the UK 72 million tonnes. It is accepted by many working in the environmental field that by 2050 the consumption should be only 25 per cent of current use (known as Factor 4), and some even argue that it should only be 5 per cent (Factor 20). A significant contribution to this target can be achieved in the construction industry by significantly reducing waste, both at the manufacturing and construction stages, and developing ways of recycling the materials used in buildings and infrastructure when coming to the end of their useful life.

There are also legal implications for anyone involved in the production or handling of construction and demolition waste, as there is a duty of care under Section 34 of the Environmental Protection Act 1990. This also includes those who are contracted to dispose of or recover waste. The contractor has a duty of care to ensure that no unauthorised handling or disposal of waste occurs. Further, in transferring the waste to an authorised person, this must be accompanied by a written description of the waste. An authorised person is usually a waste collection authority, the holder of a waste management licence, or a registered waste carrier, all of whom, in England and Wales, will be registered with the Environment Agency which issues them with a certificate of registration. The procedure recommended by the Environment Agency – adapted Environment Agency(2003) – is to:

- Check the registration certificate of the waste carrier.
- Ask where they are taking the waste to and check that the destination is authorised to take it.
- Ensure that both the contractor and the waste carrier have signed the waste transfer note.
- Keep a copy of the signed waste transfer note; they need to be retained for two years and in the case of certain hazardous materials, for three.
- Produce a description of the waste and state an accurate six-figure waste classification code obtainable from the European Waste Catalogue. Chapter 17 of the catalogue is concerned with construction and demolition wastes (including excavated soils from contaminated ground). Examples of the classification include concrete (17 01 01) and bricks (17 01 02). There are also classifications for mixtures of materials, such as 17 01 07 which is for concrete, bricks, tiles and ceramics not containing dangerous substances.
- If the wastes are hazardous, the contractor has extra legal responsibilities and may have to complete detailed waste consignment notes.
- The contractor has a responsibility to be alert for any evidence or suspicion of illegal disposal of waste, or using an unauthorised disposal site and in this case should not given the waste material to this disposer to handle. The Environment Agency should be advised.

5.2 The cost of waste

The true cost of waste is not just the replacement cost or the purchase price of the excess material. It includes the administrative cost of placing the order, progress chasing, checking on arrival and arranging payment. There are further costs in managing the waste disposal and monitoring it is done correctly and there may be an insurance liability implication affecting premium costs. The cost of landfill is based per skip rather than its composition. Since 60–70 per cent of waste is actually air voids the less waste sent, the better. Figure 5.1 demonstrates the flow of materials from its material source until returned back as landfill, but does not indicate the return flow of recycling or reuse. There are many stages were costs can be incurred such as transporting the material to site from the manufacturer or extraction supplier, storage, processing on site, removing from site, landfill

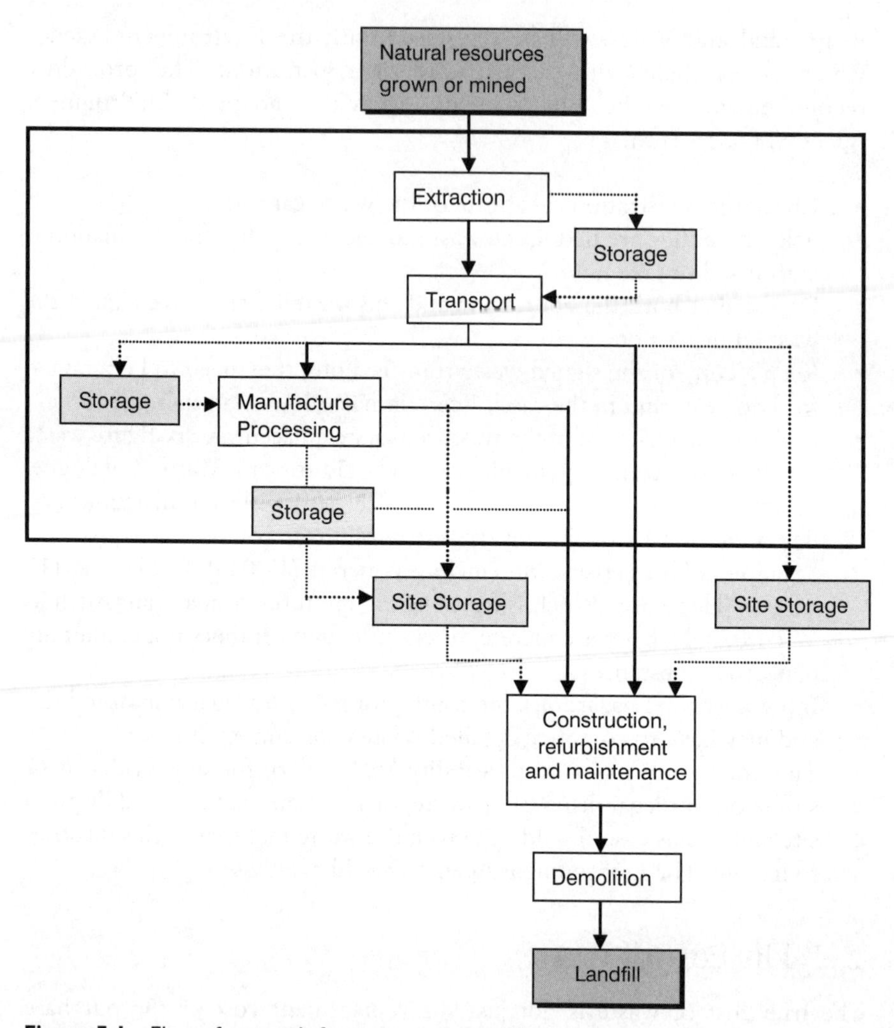

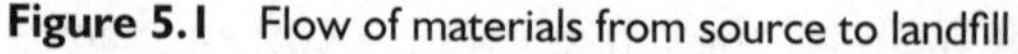
Figure 5.1 Flow of materials from source to landfill

charges as well as the potential for reuse or recycling The latter two can occur at any of the stages of extraction, manufacture, construction and demolition. The activities contained in the highlighted box are concerned purely with the extraction and manufacturing processes.

There are also environmental implications in waste disposal directly related to the extraction, manufacture or processing operations which can be reduced if the correct amount of material is ordered and used. These include the raw material, ancillary materials necessary for the process, wear and tear on plant, consumables used in the office, wear and tear on protective clothing, any packaging for protection during transportation, energy and water. There is an environmental impact from emissions to the atmosphere and effluent treatment resulting from the extraction and manufacturing processes.

Finally, there is the impact on the company's reputation. This can be as a result of not carrying out disposal correctly or the visual impact of seeing so much waste on the site making the site look untidy and potentially unsafe. It can also affect the morale of the employees on the site, especially if the waste is excessive and clearly unnecessary.

5.3 Defining waste

Waste is defined in Article 1(a) of the Waste Framework Directive as 'any substance or object in the categories set out in Annex 1 which the holder discards or intends to, or is required to discard'. Waste is anything that is discarded, even if it has value, until such time as it has been fully recovered, after which it ceases to come under the rules of the Directive. This deals with safe disposal of waste that exists and continues to occur. However, the fundamental question to be addressed is why not try to eliminate waste at every stage of the process from extraction to demolition. It is outside the remit of this text to consider the stages prior to construction, although designers and contractors occasionally can have an influence, especially on the manufacturing process. Figure 5.1 shows that materials can be sourced directly from the extraction stage, notably sand and aggregates, or from manufacturers and processors. These products can either be delivered directly as required in the construction process – known as 'just in time' – or they can be brought and stored on site prior to use. For the purposes of this discussion, the latter category can also include the intermediate stage of purchasing from builders' merchants where materials, often bought in bulk, are broken down into smaller units or packages. In these instances, the material has to be unloaded, stored and then redistributed to the place of work on the site. This increases the risk of damage.

5.4 Causes of waste

Having established the reasons and need to control waste, it is necessary to categorise the main causes of waste for analysing its material content and how this should be dealt with. Figure 5.2 divides the waste into three distinct categories. It is not drawn to scale, as the amounts would vary from site to site depending on the materials being used and the quality of the waste management. In describing these, there are occasions where some waste could be assigned to more than one category.

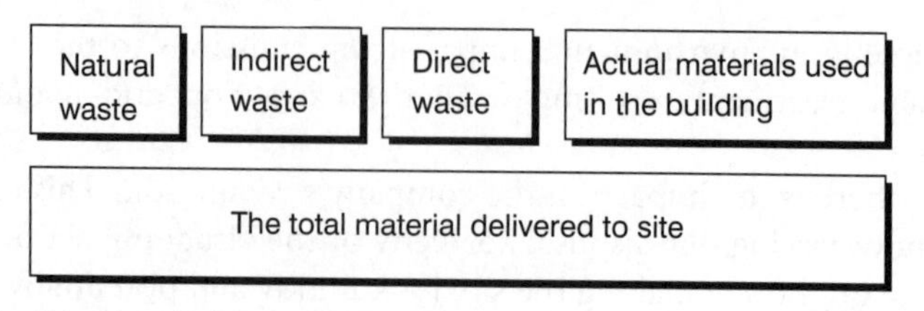

Figure 5.2 Waste on building sites (adapted from Skoyles and Skoyles 1987)

5.4.1 Natural waste

There will always be some waste that is unavoidable. Examples of this include waste resulting from cutting bricks to bond, residue left in a can, small quantities left over from a batch size, cutting out a shape from plywood or making a hole in an element of the building for a service pipe. Precisely when natural waste becomes salvageable is a mute point and difficult to give a definitive answer. Some waste is impractical to reuse or recycle, but other waste could be, except that the costs of doing so could be greater than the value of the material saved and the total costs of disposal. Where the line is drawn depends on the economic climate of the country at the time. However, this decision should be made based on calculations rather than assumption. It would normally be taken account of in the tender price.

5.4.2 Direct waste

This is waste that can be prevented with good management, planning and foresight. It occurs as a result of materials that have been lost on the site, such as bricks submerged in the mud, having to be removed from site because they have been over-ordered, or being stacked out of sight and forgotten about, in which case they have to be reordered.

Typical examples of direct waste include:

- *Transportation waste.* Every time materials are transported, loaded or unloaded there is a possibility of breakage. There is greater risk when transportation occurs on site because loads are not necessarily as well packaged and the terrain is more uneven.
- *Site storage.* Damage occurs as a result of poor preparation of the storage areas or bad stacking. Inclement weather can degrade materials if not properly protected and certain materials with limited shelf lives, such as bagged cement, can become unusable if not used within a stated period.

- *Conversion waste and cutting waste.* The dimensions of many construction materials do not match the actual size needed in the building. These include timber and plasterboard sheets, roofing sheets, timber, plastic and metal tubes, reinforcement steel bars and meshes, rolls such as plastic sheet and felts, drainage products and timber. Waste occurs as a result of cutting material to size, bond and into irregular shapes.
- *Fixing waste.* During the construction process, loss will occur as a result of errors made by the operative, such as damage caused by dropping material or a component, and incorrect fixing requiring remedial work entailing partial or complete replacement.
- *Application and residue waste.* This includes materials left in unsealed containers, materials dropped or spilt, and others, such as plaster and mortar, left over at the end of the day and allowed to set.
- *Waste due to uneconomic use of plant.* Ideally plant and equipment should be used continuously throughout the working day. Any time it is left idle or with the engine left running when not in use, is a waste. Plant should be selected to be 'fit for purpose' and not be over- or undersized for the job.
- *Management waste.* These are losses arising from poor organisation or lack of supervision resulting from bad or incorrect decisions.
- *Waste caused by other trades.* The construction industry requires many different trades to follow each other. Sometimes following trades cause damage to the previous trades' work through lack of care or understanding. This can require remedial work to be carried out. It can also include damage caused by others such as driving dumper trucks over kerbstones or grazing concrete columns when manoeuvring.
- *Criminal waste.* Any materials stolen or vandalised have to be replaced.
- *Waste due to wrong use.* It is not uncommon to observe incorrect materials being used. Sometimes this is acceptable, such as a steel reinforcement fixer using an off-cut left at the end of a cutting schedule of a larger diameter than required, because otherwise it would be wasted. On the other hand using bricks in lieu of blocks to avoid cutting is a waste of the bricks although it may have advantages from a productivity point of view. Using materials of too high a quality is also wasteful; materials of too low quality will have to be replaced.
- *Waste stemming from materials wrongly specified.* Standard specifications may have been used in the bills of quantities without analysing the required quality resulting in over-specification. This can mean either using too high quality materials or expecting too high a level of quality. For example, the quality of finish does not have to be as high if it is to be covered up, provided all other considerations are met, such as structural needs.

- *Learning waste.* This occurs when trainees are learning, not properly qualified tradesmen are used, or qualified tradesmen are carrying out operations for the first time.

5.4.3 Indirect waste

This is the difference between the estimated and actual costs incurred that cannot be passed onto the client just because it costs more. Examples of these include:

- Substitution of another material. This can occur in several ways. Where facing bricks that have been over-ordered are used in lieu of commons. Using bricks to save cutting blocks to make up a bond at a perpendicular end such as at a door entrance or butting up to another wall.
- Use of materials in excess of quantities allowed for in the specification, or on the drawings, such as the size of the bucket determining the foundation width, or uneven surfaces having to be plastered.
- Builders' errors, such as over-digging foundation depths and widths, and incorrect setting out.
- Returning to complete unfinished work.

5.5 Waste arising outside the contractor's organisation

5.5.1 The design stage

The design team can contribute to waste in various ways. This is not surprising as this is not generally part of their normal education and training and is not meant as a criticism. Examples of how waste is created are:

- Using dimensions that are not compatible with machine, material or component sizes. In the former case such as selecting a strip foundation width different to the excavator bucket sizes available and in the latter selecting the centres of ceiling joist not compatible with standard plasterboard sheets.
- Using specifications higher than are required such as the quality of the finish of concrete surfaces that are to be covered, over-designing reinforcement in a structural element, and requiring complex details that in practice will never be noticed by the users or general public.

- Designing solutions difficult to construct and are more costly than the design advantages gained. This is, of course, a subjective judgement and difficult to quantify.
- Making variations to the design as the building is being constructed causes interruptions to the works and flow of production, and often extra or different materials are required. Sometimes work done has to be dismantled to accommodate the changes.
- A lack of understanding on how the different trades are co-ordinated to carry out their work can mean them retuning to the work several times to complete it. A more complete understanding could mean the design is carried out in such a way as to permit completions of the works with fewer visits and less likelihood of damage caused by other trades.
- The design should take account of potential vandalism and materials and layouts selected to reduce the chances of this occurring. Examples of this are avoiding the use of copper pipes and lightning conductors on the exterior of a building at ground level unless protected, not providing easy access onto roofs by conveniently positioned downpipes, and not using light gauge cladding materials at ground level to deter ram raiders from entering.

The clients also cause waste by changing their minds as the project progresses, placing emphasis on unnecessary high standards for the purpose of the building, or specifying lower standards, which serves the function of the building in the first instance, but results in heavy maintenance costs and replacement of materials and components at more frequent intervals.

5.5.2 Building materials and components

Materials manufacturers still do not seem to take account of the reality of construction in deciding the dimensions they make their components. The common example of this is plasterboard sheeting, which comes in 2.4m × 1.2m boards. Yet the floor to ceiling height in the majority of low-rise residential buildings is no more than 2.3m, which means a minimum wastage of 12.5 per cent. Bovis Lend Lease (2001) estimated that 20 to 30 per cent of all London construction site waste is plaster board. On large contracts, contractors have now been able to insist the sheets come in the lengths they require, because they have the purchasing power, but smaller builders still rely on the local builders' merchants for their supply. To change the width of boards is more problematic for the manufacturer, but variations in the length are relatively easy.

Rigid and semi-rigid pipes come in standard sizes. In many cases this is not a major problem because the pipes can be jointed and much of the waste material used, although a wider range of lengths would reduce the number of connectors and hence reduce waste. Pipes such as rainwater goods that are seen, need to be jointed as infrequently as possible for aesthetic reasons, but the current range of pipes sizes is limited and still results in significant off-cuts. The BRE recommended as long ago as the 1980s that manufacturers should produce half and quarter sizes, but even a cursory look at major manufacturers' web sites demonstrates that little notice of this request has been taken. There is also a similar problem with sawn timber.

The design of packaging used for protection of goods during transit and to permit bulk handling at the destination is another source of potential waste. Bulk packaging such as used for bricks, blocks and pipes inevitably means that waste will occur as the builder may not require the total content of the package. The percentage waste where large quantities of these materials are being used is small, but on smaller orders this increases. On the other hand the advantages of this form of packaging from a production point of view are self-evident, because of the speed of unloading, the reduction of labour resource needed and the ability to deliver on time and with more flexibility, but it has been suggested that smaller packages of materials could also be accommodated.

The trade literature provided by some manufacturers does not always include the full range of sizes available. This means purchasers will tend to select from those shown rather than investigate further and reduce waste as a result.

The materials used in packaging can be excessive, as is well-documented in the retail trade, often used to make the purchase more attractive to consumers. There should be no need for this with building products, except perhaps in the DIY stores for certain items, although this is debatable. Some of the packaging such as strapping and cartons can be recycled but not re-used, but dunnage and pallets are usually re-usable. Some 15 to 20 per cent of all construction waste is packaging. The problem is how best to return this waste to the supplier.

Wastage also occurs as a result of poor packaging, loading, stacking or unloading by the supplier. Not only is there waste material, but it can also affect production if the materials are 'just in time'. This is important to note, that as the industry moves in this direction, it is essential all goods arrive intact.

The contractor must be advised by the supplier of any special requirements for the unloading of the materials or components. This may be where the lifting points are, the lifting equipment needed, the way it should be stacked or stored, and the direction the product or container should be kept. There should be a clear understanding between the site and the supplier on the sequence that products are required so they are packaged correctly. For example, precast concrete or steel structural components should be loaded so they can be taken straight from the vehicle into the building without causing any undue stresses on the vehicle due to the load becoming unevenly distributed.

5.5.3 Plant sizing

There is a conflict of interests between the needs of an individual construction site and that of the plant manufacturer. The former has specific issues to be addressed, and the latter has to produce plant that will satisfy a wide range of customers whose construction operations are not identical. Examples of this include excavator bucket sizes that are not compatible with the width of trenches resulting in producing wider than needed trenches, cranes with over-capacity, or loads having to be split for lifting purposes, and a limitation in the range of tasks for which the plant can be used.

Other plant may be limited in its manoeuvrability, for example not being able to easily steer round obstacles, or having inadequate suspension to cope with the uneven construction site terrain resulting in spillage, damaged material or slower operation, because of the caution needed.

5.5.4 Communications

In the issues discussed above, it becomes clear that good communications between all stakeholders is needed. Further, if each understands the other's needs and limitations, this will improve the situation.

5.6 Construction site waste

There has been a significant change of attitude within the construction industry in recent years to the disposal of waste due to increasing awareness of environmental issues and the introduction of landfill taxes which have increased four-fold since introduction and will almost certainly continue to rise.

Waste can be divided into three categories that can then be subdivided further. The three types are:

- *Inactive waste.* These are materials which do not undergo any significant physical, chemical or biological reactions or cause environmental pollution when deposited in a landfill site under normal conditions. The kinds of construction materials that fall into this category include soils, rocks, concrete, ceramics, masonry and brick and minerals.

- *Active waste.* These undergo change when deposited. These include acids, alkaline solutions, pesticides, wood preservatives, oils, asbestos, timber, plastics, bitumen and batteries. Active waste attracts a higher landfill tax than inactive.
- *Special wastes.* These are classified as being deemed to be dangerous to life. Some of these overlap with substances classified as active waste. It depends on the composition and relative risk and is covered in the Special Waste Regulations 1996. They are materials that are toxic, very toxic, corrosive, reactive, explosive, carcinogenic or flammable. Typical examples include acid and alkaline solutions, waste oils and sludges, and wood preservatives.

5.7 Waste recycling and re-use

Materials should be segregated at the place of work. To attempt to do it at the collection points is difficult once the materials have been mixed because of the nature of the materials especially those used in wet trades. The categories of material normally segregated from new build operations are the inert materials such as concrete, brick, block, timber, paper and cardboard, metals and some plastics. Soils, ceramics, plaster, insulation materials, bitumen, cans and containers used for toxic materials will usually end up in some form of landfill. In maintenance and refurbishment, other materials will be involved, notably asbestos which will have to be dealt with carefully, the method determined by the category of the fibres.

To segregate on site requires space for the bins to collect the materials, methods for getting the material from upper storeys to the bins, usually chutes, and methods of transporting the materials on the horizontal plane. Finally, systems have to be in place to ensure the segregation process on site actually takes place bearing in mind that the majority of operatives will be sub-contractors. This can be accomplished either by having a gang directly employed by the main contractor, or as part of the contractual arrangements with the sub-contractor. The advantages of segregation on site is it is easier to see what types of waste are being generated, monitor the quantities and leading to focusing on where waste reductions should be targeted.

However, all of this fails if there is nobody locally who wishes to collect and recycle the materials. Most contractors will have the names and addresses of companies who will do this, and the local Environmental Health Office can usually help. Some businesses advertise on the web.

Table 5.1 UK construction waste (source BRE)

Materials	*Percentage*
Packaging	25.0
Timber	13.8
Plaster and cement	11.5
Concrete	10.2
Miscellaneous	9.6
Ceramic	8.6
Insulation	7.5
Inert	7.1
Metal	4.0
Plastic	3.2

There is a move towards leasing the material for the duration of its life then collecting and recycling it when the end of its life is reached. Two examples of this are road surfacing materials and carpets. The main materials where there is a proven market for recycling are concrete, blacktop, topsoil, excavation spoil, timber, metals, bricks, stones and pipes. There are also specialist markets for reusing architectural features such as fireplaces, rainwater goods, plaster covings and ceiling roses. Plastics is a developing market and some packaging can be reused providing the supplier wishes to participate. A word of caution with timber: increasingly timber used in buildings has had a preservative treatment, using chemicals that are very toxic and not always obvious, so reuse of these materials, especially if previously structural, needs to be carefully thought out. Burning for fuel could result in toxic emissions.

Reductions in waste can be achieved by analysing the amount of waste being produced, determining the cause and putting in place an action strategy to reduce or eliminate it. For example, setting waste targets, changing the method of transportation and packaging, arranging for materials to be pre-cut before delivery to site and modifying a design. Table 5.1 gives an indication of the breakdown of construction waste.

5.8 Implementing a waste minimisation policy

A waste minimisation programme comprises various stages and starts by concentrating on attitude and the culture of the organisation. As with any company policy, to be successful, all members of the organisation have to eventually engage in the process. This can only occur if the senior management are committed to making it happen. Whilst individual site managers can have an impact, without the support from above, it is unlikely they will be as successful as they might be. This is because to minimise waste other key players, such as the purchasing department, have to play their part.

To engage and convince the senior staff, they will need to see the data that demonstrate the implications and cost savings that will be made, and contrary to popular belief, it would be rare if savings were not possible. Once convinced and committed, senior management will have to commit resources, usually people's time, to the various stages of the process.

The next stage is to engage the staff in the process. There may be some reluctance to change (*Business Organisation for Construction*, Chapter 7), as it can appear as yet another senior management initiative requiring more time and effort on top of an already time-pressured employee. Further, if asked to look for areas of wastage, if found, they can be afraid they will be criticised for not dealing with it earlier. Using several methods including posters, newsletters, team briefings, workshops, information enclosed with pay slips, and suggestion schemes can raise awareness.

It is not unusual for waste to be considered simply as the material left after being fashioned, but by stepping back and investigating the whole process it can be seen to have wider implications, as shown in Figure 5.3 showing a flow diagram of the production process and Figure 5.4 showing the flow diagram of construction of a one-brick-thick brick wall, assuming no scaffolding or plasticisers are required.

If the process is then changed to that of an external cavity wall, the basic materials would additionally include, wall ties, blocks and insulation material. The waste stream would also include strapping from block packs, block off-cuts from bonding, damaged blocks, blocks surplus to requirements, insulation off-cuts and insulation surplus to requirements. If the wall included openings then the use of lintels and formers to form the opening would be added.

An analysis of the causes of waste is necessary to establish whether the waste is as a result of the supplier (standard size components and materials, etc.), the type of packaging, or the production process including how the materials are handled and stored on site, so suggestions on how this can be improved is thought about and action taken prior to production commencing.

The collection of data can be divided into two parts: theoretical waste and actual waste. The former can be established by taking the data from the estimate, the amount of wastage allowed for and the quantities requested on the orders. From this can be calculated the theoretical waste for all the

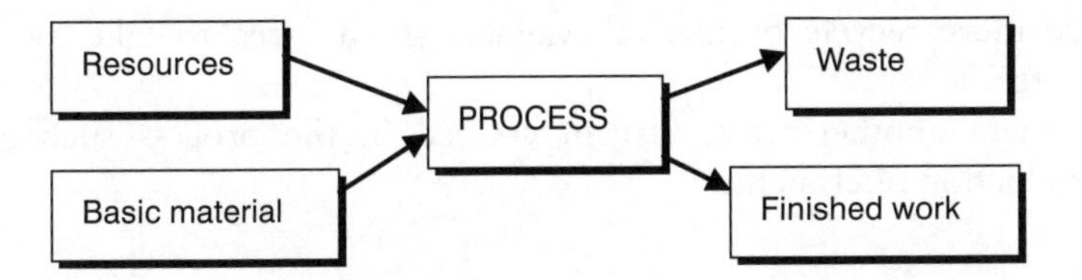

Figure 5.3 Flow diagram of production process

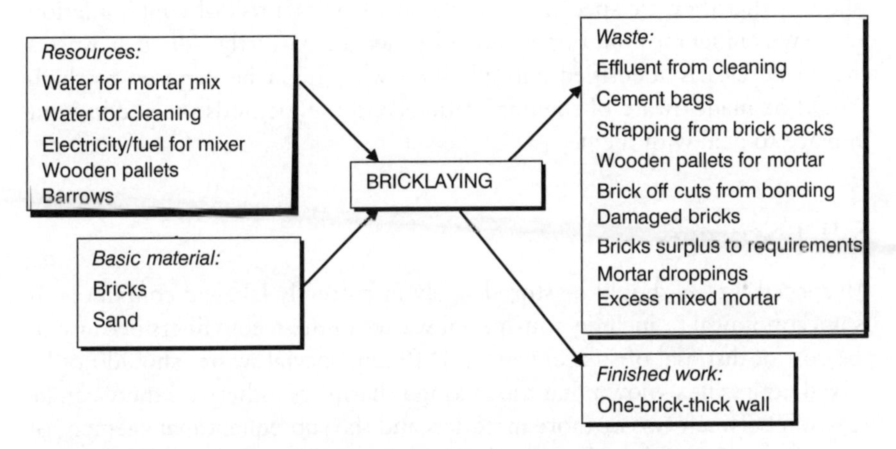

Figure 5.4 Flow diagram of the production of a one-brick-thick wall

basic materials. The latter can be established by observing what happens to the waste stream and its content. If these are not the same, analysis should be undertaken. However, if similar, complacency must not be allowed to set in. The next stage is to set realistic targets for improvement and continue the monitoring process with the aim of eliminating waste as far as is practicable. Note should be taken of any extra ordering that is required in excess of the initial order.

Part of this process is also to prioritise which materials should be targeted in reducing waste and then recycling at disposal. Questions that should be asked include:

- Which waste is the most expensive to replace?
- Which waste has the greatest recycle value?
- Which waste is the easiest to collect?
- Which on disposal has the greatest environment pollutant effect?
- Is it a process that causes a nuisance to others?
- Is the best method of work being utilised?
- Are there obvious ways of reducing waste?
- How easy is it to segregate the waste?

- Are there recycle businesses available in the area to take away the waste?
- Is there another waste stream created in the process such as the production of effluent?

5.9 Disposal of special waste

The fact that they are special wastes means there is a risk of contamination and environmental pollution if not disposed of correctly. All such wastes should be clearly identified and all those who might be exposed to them should be made aware of the implications and the methods the site has put in place to deal with them.

5.9.1 Storage

All special wastes should be stored safely in correctly labelled containers. It is uneconomical to include non-special wastes in these containers because of the cost of disposal of special wastes. Different special wastes should not be mixed, unless it is known that the mixing is harmless, otherwise there can be reaction between two or more materials and the subsequent management of the waste at disposal made more difficult.

5.9.2 Transport and disposal

The local environmental agencies have to be informed in advance using a special waste consignment note if special waste is to be disposed of. This note must also accompany the transport of the special waste in lieu of the usual waste transfer note (section 5.1). An appropriate disposal route has to be identified so as to reduce risk in the event of an accident en route and the documentation has to be stored safely.

References

Biffa (2002) *Future Perfect*. Biffa.

Coventry, S. and Woolveridge, C. (1999) *Environmental Good Practice on Site*. CIRIA.

Curwell, S., Fox, R., Greenberg M. and March, C. (2002) *Hazardous Building Materials: A Guide to the Selection of ,Environmentally Responsible Alternatives*, 2nd edn. Spon Press.

Environment Agency (2003) *Construction and Demolition Waste: Your Legal Duty of Care*. Environment Agency.

Environment Agency and The Natural Step (2002) *Towards the Sustainable Use of Materials Resources,* 2020 Vision Series No 4. Environment Agency.

European Economic Community (1991) The Waste Framework Directive (75/442/EEC as amended by 91/156/EEC). Available: faolex.fao.org/docs/texts/eur38116.doc. Accessed 14 October 2008

Skoyles, E.R., and Skoyles J.R. (1987) *Waste Prevention on Site*. Mitchell.

CHAPTER

Stock control and materials management

6.1 Introduction

This chapter can be read in conjunction with Chapter 7 in this book and Chapter 12 in *Finance and Control for Construction*. When a high percentage of the site construction work was carried out by the main contractor using directly employed labour, the control of stock, or lack of it, was a significant issue. Much of this obligation now has been transferred to the sub-contractors. The contractors still have to appreciate the implications of stock control as they have to procure materials and components for labour-only sub-contractors and they can also be the sub-contractor on a large contract. They have the responsibility to ensure the quality of materials brought onto site for use in the building is of the correct quality, as rejected material can effect progress. As the industry moves towards increased supply chain management, there is a further role in supporting sub-contractors in their control processes.

The purposes of stock control can be summed up in two simple statements:

- to control the number of items necessary to run the operation without carrying unnecessary stock; and
- to act as a buffer to insure against delay and uncertainty.

6.2 Types of materials

It is necessary to categorise materials into different types, as they require different treatments in the way they are managed and controlled. In Chapter 1, the concern was about protecting material from damage, but the emphasis is different for stock control as one of the key issues is the duration the materials are needed on site before being incorporated into the building as this has a direct impact on the date the material is ordered. Categories include:

- Raw materials, such as bricks, blocks, aggregates, cement, uncut reinforcement steel and timber which have to be processed to produce the final component of the building. In some cases they will have to be stored and then transferred to the place of work, in others, distributed to the place of work on arrival.
- Partially processed materials, such as cut and bent reinforcement steel, which has to be stored, assembled in situ or prefabricated and lifted into position, and ready-mixed concrete which is poured directly into the required location. Other examples are softwood timber window and door frames that have to be painted.
- Completed components such as uPVC windows, precast concrete cladding units, kitchen fittings, hardwood door frames, boilers and air conditioning units.
- Small items used in the construction process such as nails, screws, nuts and bolts, ironmongery, paints, adhesives and preservatives.
- Office supplies such as paper, pens, pencils, paper clips, envelopes and printer cartridges.
- Maintenance materials such as cleaning fluids, light bulbs and refuse bags.

Inevitably, some materials can fit into more than one category such as whether an off-site manufactured component is completed or partially completed on arrival on site because fixing materials are not included.

6.3 Stages of stock control

To fully appreciate the subject of stock control it is necessary to start at the manufacturing stage and to understand that manufacturers of different materials produce their output in a variety of ways which can effect the purchaser's requirements. Five different examples:

- Rolled steel is a continuous process that only stops when the rollers controlling the size of the section wear out and reach the outer limit of size tolerance. At this point the rollers are changed to new ones either to continue making that section or start another. If steel is in short supply and little is in stock, it may be some time before the size required is made again.

- Precast concrete components are not made in the sequence they are to be erected on site, but in runs of the same type of component because to change the mould to another type may mean a day or more in lost production. On completion the components will have to be stored in the yard until they are properly cured. One of the effects of this is that if a component is damaged on site it cannot always be replaced immediately unless the supplier has one of the same types in stock.
- Kitchen units, on the other hand, are largely made from the same size component parts, the only difference being the facings such as drawer fronts and cupboard doors, so the supplier can react more rapidly.
- Bricks and block manufacturers are continually making the same products and therefore the only problem likely to be met in replacement or increasing an order is if stocks are very low because of demand elsewhere. However, special bricks will take longer to replace as it unlikely they are in stock.
- Plant and equipment for incorporation into the building often is purpose-made and sometimes designed by the supplier. Depending on the complexity of the item, this process can take several months to complete.

Manufacturers require different lengths of time from the order being placed until they can deliver to site, known as the lead-time, and many not be able to react rapidly in the case of design changes or replacements. It should be remembered, except in certain circumstances, the manufacturer is servicing many other clients some of whom may be more important to them in the long term than others.

Once on site a further period of time may have to be added depending on what processing has to be carried out before fixing into the building or to build up a stock of materials or components before this occurs. For example, a section of the building may comprise several large elements each individually off-loaded by a mobile crane brought in for this purpose. When all components have been delivered a larger and very expensive crane is brought in to erect the components that may only take a short time to accomplish.

Bearing these points in mind, the quantity of materials required for the contract, when they are required and the quality and specification of the goods can be established. The selection of the suppliers is covered in *Finance and Control for Construction*, Chapter 12.

6.4 Problems of excessive stock

Whilst the construction industry is shielded to a certain extent by most contractual documents, the client paying for goods on site even though not incorporated into the building, there are still many reasons why excessive stock is a problem and should be controlled, including:

- excessive stocks take up capital which could be better used, and in spite of the shield mentioned above, most small items in the stores, such as nails and screws are rarely costed in the monthly valuations;
- they take up space on the site and absorb further capital for the supports and protection against the weather, which has to be provided;
- they are more likely to be damaged resulting in unnecessary expenditure and possible delays incurred in replacement;
- it may be necessary to double handle stock to elsewhere on the site and missing the opportunity to have just in time delivery when the components go directly into the building.

6.5 The storage function

This includes all materials arriving on site that are stored or, in some cases, for goods delivered just in time. The value of the materials incorporated into a building represents anything from 40 to 60 per cent of the final cost of the building, so tight control is essential. This involves receipt of the goods, inspection, issuing, stocktaking, dealing with replacements and disposal of damaged or excess goods. There is a danger on construction sites that this responsibility gets shared between various persons; the secure store accommodation might be in the control of a storekeeper, who will also be responsible for receiving goods, but the storage areas around the site are the responsibility of another. In the latter case, that person will have many other duties besides looking after stored goods with the result they are not looked after and controlled as they should be.

6.5.1 Receiving

There is certain information the receiver of goods must have to carry out their function properly. They need a copy of the order indicating the quantity and types of goods, the date of delivery and, in the case of scheduled deliveries, all the dates when goods are planned to arrive. They should be furnished with information as to where the materials are to be stored or, in the case of just in time deliveries, where the materials are to be sent and any special advice and instructions on checking procedures for a particular load. This information gives the storekeeper the knowledge to plan for space and the methods of storage such as shelving and bins in the internal sheltered secure stores.

When goods arrive they need to be checked against the order to ensure everything has arrived as expected and is undamaged and of the correct quality. If for any reason it does not comply, action has to be taken to rectify the situation as production could be held up as a result. There needs to be clear procedures as to what must be done and at all times records kept so the story can be tracked at a later stage if necessary. Decisions such as if and when damaged goods are returned to the supplier, who notifies the supplier of this or other discrepancies, should have been decided when setting up the site. It should be noted that the delivery ticket may not tally with the order requirements and this should be checked. The date and quantity of accepted goods received should be recorded in such a way as to allow the quantity surveyor to calculate the value of the goods received on site for the monthly valuation. With the development of computer software, this process has been simplified.

6.5.2 Inspection

It is not possible to check every item on every lorry as it arrives as in many cases it would take too long or mean unpacking items to inspect the contents, so procedures have to be in place to account for this. There are four ways checking can be used that will cover most eventualities:

- Experienced staff carry out visual inspections. The condition of materials delivered in bulk, such as bricks, aggregates and sand can readily seen to be of the right standard. Bricks and blocks will show signs of damage and if sand has excessive moisture content, water will be seen coming out of the tailgate.
- The complete contents of the load are inspected if practical. These are usually components such as precast concrete, structural steel, window and door frames, manufactured cladding components and kitchen units. These usually are readily visible.
- A random selection of smaller items can be taken and checked, and if these are correct, the assumption can be made that the rest are also in order and the load accepted as per the delivery note.
- Deliveries can be tested on arrival. Samples of ready-mixed concrete can be taken and tested for slump before pouring and cubes made for analysis

later. Bulk loads of sands and aggregates can be sent to a weighbridge to check that the load is as heavy as stated.

6.5.3 Storage and location

This is discussed in materials storage and handing (section 1.6), which looks at types of materials and the criteria for storage. The key issues are that material should be stored correctly to stop damage by the elements, third parties, contamination, distortion or breakages; it should be made secure; and the flow of materials out of storage should be on the principle that the oldest in stock should be the next to be issued.

6.5.4 Issuing

It is essential there is strict control of the issuing of materials. Clearly how much control is put in place depends on the type of goods. If a cladding component is being lifted from a storage area to be fixed to the façade of the building, then it is not necessary to have an issuing procedure as the panel has only one purpose and can be identified easily. At the other extreme, the issuing of items such as door furniture needs to be carefully monitored as these are easily stolen. This is relatively easy to control providing goods are properly issued. The danger is that employees go into the store and take what they need without it being booked out properly.

The procedure starts with the request for goods to be authorised by the supervisor. The employee can then go to the stores, receive the goods issued against the requisition, and have the date and amount recorded by the storekeeper. This means the storekeeper has a 'paper' stock count at any point in time and if stocks are dropping below a pre-defined level, can call up further materials.

On a construction site, the most difficult items to monitor are materials such as timber that can be used in many different locations and in different sizes. Take any finished room and inspect the different lengths of skirting that have been fixed. Remembering the material arrived in standard lengths, imagine how to control the issuing of the correct amount required to complete the task so that minimal waste occurs. It is very easy for wastage to occur, unless consideration is given to a cutting schedule to obtain the optimum use of the timber. This is standard practice for reinforcement steel. The use of these materials should be monitored regularly and wastage measured and compared against set standards.

6.5.5 Stock control

There are three aspects of stock control. First, to ensure there is always enough in stock to support the production process; second, to test the systems and procedures used to control stock are working; and third, to ascertain the value of goods being held. Part of this process is to check if goods are deteriorating in stock by inspecting shelf life dates, the state of goods, such as rusty nails in the store and in outside storage areas, the general condition of bulk materials and factory made components. Where deterioration is found, steps need to be taken to eliminate this in the future and to re-order replacement goods and materials.

There are a variety of reasons for having stock on site where it is not practical to have just in time delivery. By buying larger quantities at a time it may be possible to negotiate a lower price by having full loads delivered. It can also reduce the administrative costs of purchasing, resulting from fewer orders being placed, receiving the goods, checking and handling.

6.6 Just in time deliveries (JIT)

Much has been written recently about JIT deliveries. Much of the original work was carried out by Toyota, and because of the success, the rest of the world has followed suit. However, it is not a new concept, nor is it new to the construction industry. In the 1960s many of the industrialised prefabrication systems relied on components being delivered to site and incorporated into the construction on arrival or within a few hours. The majority of concrete is also delivered on the JIT principle. It is debatable whether or not the construction industry could go totally to JIT, because of uncontrollable factors when working outside, like the weather and finding the unexpected when working in the ground, but once the building is water tight, then this is another matter.

The logic behind the principle is as follows: the reason for holding stock is to cover for short-term variations in demand and supply. These stocks serve no useful purpose and are only in existence because of poor co-ordination of materials by management. If stocks are allowed to remain, there is no incentive for management to improve, resulting in the problems being hidden by the stocks. Therefore the best option is to improve management to find the reasons why there are differences between supply and demand and take the necessary action to deal with it. The result of this is to demonstrate that JIT is not just about delivering materials just in time, but also to improve management as a whole.

On further investigation, besides reducing or eliminating stock held, other benefits can accrue. Normally, if a piece of plant breaks down, production is reallocated elsewhere and incoming materials are stored until the plant is repaired. The question to be answered is why should the plant break down in the first place, because if it didn't, there would be no need to make these changes and production would remain at a higher level. There is often some suspicion between the contractor and suppliers often based on the experiences of both. JIT means the contractor has to totally rely on the supplier and this means building up trust and co-operation. The same co-operation is required between the employees and management to eliminate friction between the two parties. Finally, extra stock is sometimes kept for the eventuality of defective materials. JIT encourages there to be no defective materials, by working closely with the supplier to ensure this cannot occur.

The principles of JIT appear very simple, but of course it is not so easy to implement in practice, as it requires a change in attitude. There needs to be in place some key elements if it is to be used successfully. These include:

- Maintaining continuity of production without switching from one operation to another otherwise the planned flow of materials will be interrupted.
- Encouraging the standardisation of components as this enables JIT to work more effectively. The more variations the more difficult is the control required, increasing the likelihood of errors.
- To control delivery costs, the work should be packaged in such as way as to use full loads on delivery if possible. If a partial load, the cost per item increases and there is a loss of control in delivery time as the vehicle has to go to other locations.
- Careful consideration to the distance and complexity of the journey over which goods have to be supplied. Usually, the longer the distance, the greater the risk of delay. This needs to be built into the programming of deliveries.
- Suppliers have to ensure that goods arriving are as meant and are not defective in any way. This requires stringent controls at the point of loading, appropriate packaging, correct equipment available for unloading on site, and properly trained and supervised employees.

The relationship with the suppliers is paramount to the success of JIT. This can require change in attitude by both parties as contractors have traditionally looked for the lowest tender price rather than building up any long-term relationships. Suppliers have had little customer loyalty, seeking out any customer available and trying to make the highest profit. JIT requires

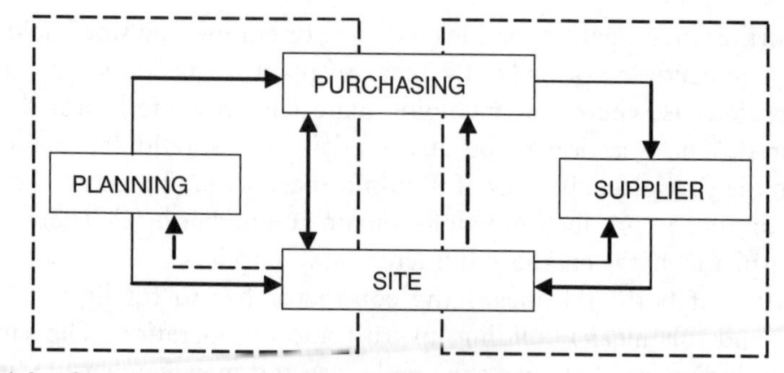

Figure 6.1 Communication: the control of materials

that both sides have the same objectives that are beneficial to both and the need to develop long-term partnerships. A word of caution though: this should not be a sole partnership whereby the supplier only has one customer. The construction industry is too volatile to give a guaranteed demand.

6.7 Communications

There needs to be a communication link between the person placing the order, the person controlling the flow of materials through the site, the projected programme requirements and the actual site production otherwise control will be lost. There are two 'circles' of communication in the process as shown in Figure 6.1. The planner, after producing the programme for the contract, advises both the purchasing department and the site so that orders can be placed and the work planned on site. The purchaser agrees with the site a schedule of deliveries after which an order can be placed with the supplier. The supplier delivers to the site, relying on more up-to-date delivery requirements from the site. If there are any problems about the reliability of service that are not amicably resolved, the site advises the purchaser so consideration can be made as to whether to place further orders in the future. At all times the site progress is monitored.

References

Waters, D. (1996) *Operations Management*. Addison-Wesley.
Womack J.P. and Jones G.T. (1996) *Lean Thinking*. Simon & Schuster.

CHAPTER

7 Supply chain management

7.1 Introduction

Supply chain management as a subject has come to prominence in construction since the Egan report *Rethinking Construction*, but has been in evidence long before, especially as a result of the work done in the Japanese motor car industry. Many academics have taken the position that the work done in manufacturing can be transposed into construction, but this should be treated with a degree of caution as the working environments are different in several significant ways. In manufacturing or retail, the factory or store is in a fixed location with a projected long life, whereas in construction the site activity usually spans between six months and two years and could be located anywhere within the area of the company's operations which may encompass large regions or all of the country. Sub-contractors who have to be physically on the site carry out a significant percentage of the construction activity. This adds a further dimension to supply chain management, as there may not be a match between where the contractor is working and that of the supplier. It may be uneconomical to use sub-contractors located at a distance from the site activity. Further, the contact conditions may require that a minimum number of employers on the site be resourced from the local community.

Suppliers of materials on the other hand are similar to those servicing the retail and manufacturing business, in that materials and components generally are stored or made in fixed locations, the materials and components being brought together for assembly. The major differences are that for some of the suppliers, the product required for each contract will be purpose made requiring considerable changes to plant and equipment at the supplier's base. Other materials and components such as timber, aggregates, reinforcement steel, structural steel, window frames and door frames will require no modifications to the supplier's processes and only minor adjustments to their plant. Delivery problems are similar other than the delivery point is continually changing as one contract closes and another commences. The activities where some of these variances can be eliminated or reduced is where there is a contractual relationship between the client and a contractor for multiple contracts, as for example with some clients in the retail sector, or where the client and the contractor are one and the same person as in speculative housing..

Construction plant for earthmoving, lifting, transportation, power provision and mixing has to be procured for the construction of the project, unlike manufacturing where the equipment will be in place. Similarly temporary works such as scaffolding are required.

There has been a reluctance by construction clients to enter into supply chain relationships with main contractors, which makes it more difficult for supply chains to be developed with sub-contractors and suppliers because of lack of guaranteed orders. Further, the client, perhaps because of the types of contracts in use, tends to communicate to sub-contractors via the main contractor thus keeping out of the supply chain management at the lower levels. At the time of writing much of the supply chain management in place is for the duration of the contract only.

This reasoning in demonstrating differences is not meant to dissuade against the argument for supply chain management, but simply to show there are differences and one cannot always transfer the practice of one industry to another and expect it to work as effectively.

7.2 What is supply chain management?

The supply chain, or supply side, comprises those organisations which supply materials, components, labour, plant and equipment to enable the site operations to be accomplished.

Supply chain management is the active engagement of management in the activities of those involved in the chain to ensure best value for the customer and to achieve a sustainable competitive advantage. This may mean supporting an organisation in the chain that does not have the resources or systems in place necessary for the needs of construction processes such as conformity to ISO 9000 (Chapter 8) and could involve some financial investment by the contractor. Above all it is about the relationships and trust between those in the supply chain and the contractor. Traditionally this has often been poor because of the emphasis placed on the conditions of contract made between the parties and the protection of personal interests rather than the common good.

It should be remembered that the contractor is part of the customer's supply chain and when the building is handed over, the customer may in turn be part of somebody else's supply chain, often referred to as the demand side. This total picture is referred to as the supply chain network.

7.3 Supply chain network

If the business looks solely at those suppliers and sub-contractors that are servicing the organisation there is the possibility these could be let down by a failure of one of their suppliers. Therefore all the links in the supply need to be considered. Figure 7.1 illustrates a typical supply chain network for a building project on the supply side. For simplicity it has ignored the supply of plant and equipment and temporary works. The term source is used as the place of extraction and processing. From this it can be seen there are several levels before those that directly impact with the site activity. There could be more levels as well. This figure shows the relationship to a specific contract, but can be developed to look at the supply chain to the company as a whole. Figure 7.2 illustrates the demand after the building has has been handed over. The third level only shows the network for the shopping tenants.

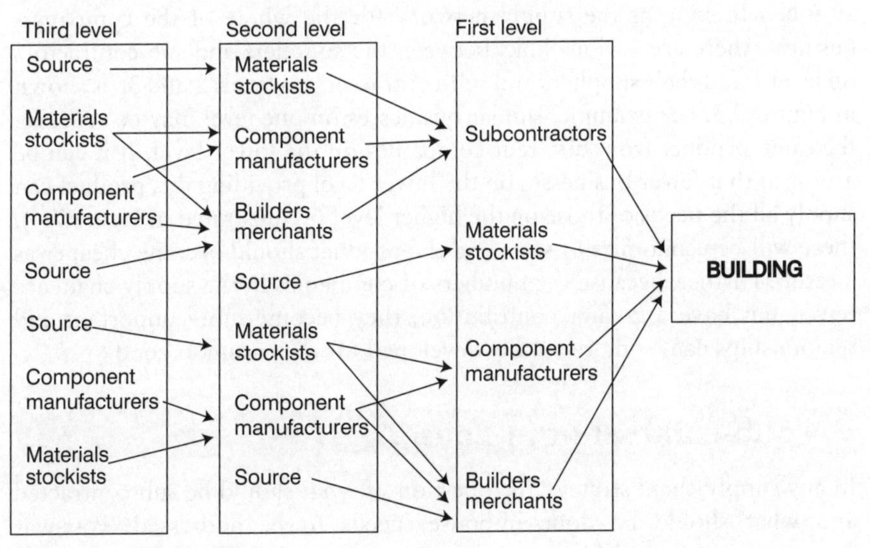

Figure 7.1 Supply side

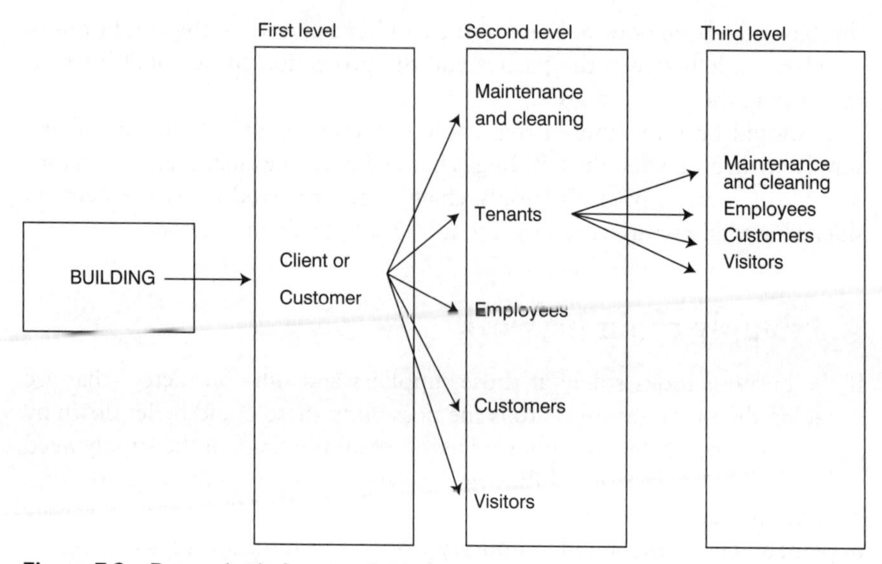

Figure 7.2 Demand side for a retail development

The advantage of producing the network is it can assist in making the company more competitive as it demonstrates the needs of the customers and highlights potential weaknesses on the supply side; it assists in identifying significant links in the network thereby focusing on where breakdowns in the link will have the most disadvantageous effect; and the long-term effect of weak leaks can be analysed with a view to strengthening or replacing them.

When looking at the supply network for the whole of the company's business, there are various links between the suppliers and sub-contractors on level 1 and their suppliers and sub-contractors on levels 2 and 3, as shown in Figure 7.3. For example, similar businesses on one level may be sourcing the same product from different companies on the lower level. If it can be arranged that fewer businesses on the lower level providing this product can supply all the needs of those on the higher level (as illustrated in Figure 7.4), there will be economies of scale and the product should become cheaper as a result. Further, because the numbers of companies in the supply chain are fewer, but have a greater contribution, they become more important and relationships can be more readily developed for the common good.

7.4 Sub-contract or in-house?

In any supply chain strategy the question of what should be sub-contracted and what should be done in-house, needs to be addressed. Post-war most contractors would directly employ joiners, bricklayers, steel fixers,

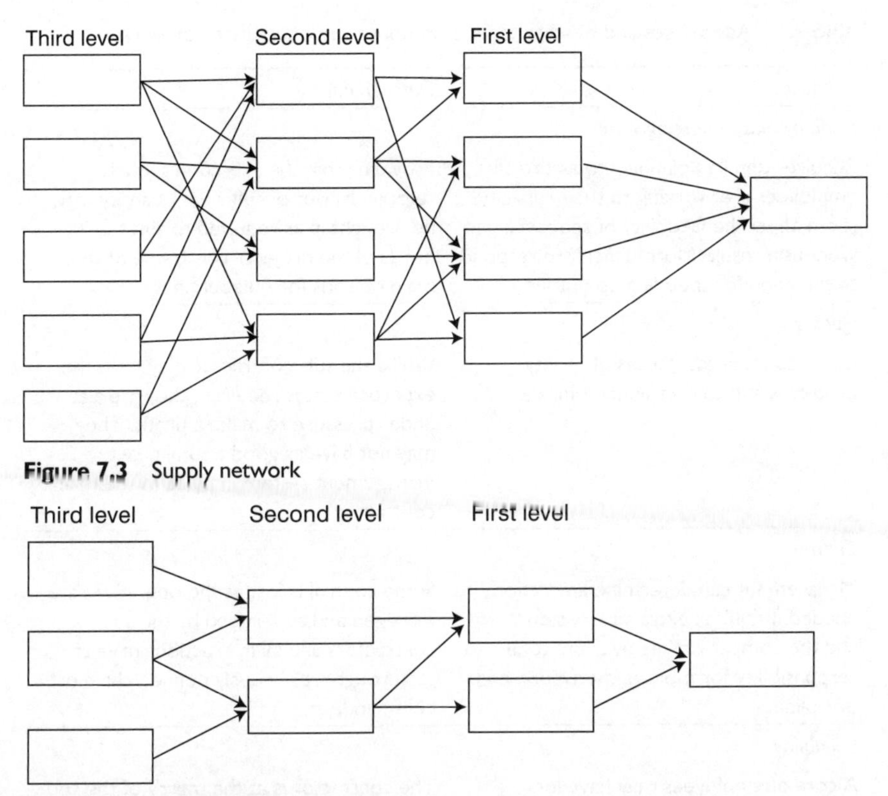

Figure 7.3 Supply network

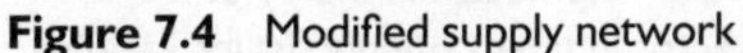

Figure 7.4 Modified supply network

scaffolders, plasters and general operatives. The only sub-contracted work would be for the mechanical and electrical services. There has been a steady trend from the 1970s to sub-contract nearly everything, keeping just a small gang of general operatives to unload, tidy up and other such tasks. There have been some pockets of resistance to this in certain parts of the UK. In recent years some of the site staff functions, notably site engineers, have also been outsourced. In the immediate future this trend is unlikely to change, but it is argued that periodically the organisation should revisit this issue, not least because of the concern that smaller sub-contractors are less likely to offer apprenticeships, so who trains for the future? On the supply side, some contractors had, as part of their history, a manufacturing arm, often in timber components, but this has generally died out. Some contractors have a mineral extraction operation, but these are generally run as separate companies and normally would be treated as any other supplier. Table 7.1 looks at some of the advantages and disadvantages of in house provision and outsourcing.

Table 7.1 Advantages and disadvantages of in-house and outsourcing of work

In-house	*Outsourcing*
Planning and resourcing work	
Requires detailed planning to assure all employees are working to their capacity. It can affect the flexibility of sequencing work as a result. More difficult to respond to the need for speeding up output.	Providing that the sub-contractor is reliable, labour of sufficient quantity can be brought in as required to suit the needs of the project. This is one of the main reasons for outsourcing.
Quality	
Easier to trace the causes of quality problems and to take action quickly.	Whilst the sub-contractor may have the expertise that is required, they are also under pressure to make a profit. They may not have as good a total quality management system in place as the main contractor.
Control	
Management can determine any action needed. Requires extra supervision and the company takes over the total responsibility for morale, motivation and discipline.	Some control is lost as the operatives are managed and supervised by the sub-contractors and all instructions have to go through a previously defined chain of command.
Reliability	
A core of employees may have long service records, know the way the company works and are compliant with its needs and are more likely to look out for its interests.	The contractor is at the mercy of the sub-contractor in terms of punctuality, keeping to programme and quality. It is in the sub-contractor's interest to perform well if new work is to be obtained. There may be some flexibility in being able to move employees from one contact to another providing there is enough critical mass.
Costs	
Any contribution the operatives make to the company profit is kept. However, when the contract is completed, there can be costs involved if the 'key' operatives are kept on whilst waiting the next contract to commence.	The other main reason for outsourcing is that the overall costs to the project are likely to be lower as sub-contractors are in competition with each other to obtain the work. The main administrative costs in terms of wages, transportation to site, accommodation, etc. is borne by the sub-contractor.

Companies engaged in design and build and other forms of procurement where they are responsible for the design element, may well contract out parts or all of this process. This is especially the case for the building services, and may include architectural and structural design if the contract is of a specialist nature and the company cannot source within this level of competence and expertise.

The decision-making process on whether to outsource or not is a series of progressive questions. Is the activity of strategic importance of the company? If not, then does the company have the specialist knowledge to carry out the work? If not, then does the company carry out this work better then its competitors? And if not, is it anticipated that in the future it will improve its importance? If the answer to this final question is no, then outsourcing should be considered. On the other hand, if the answer to any of the previous question is yes, then serious consideration should be given to keeping the activity in-house as it would probably be unwise to outsource.

7.5 Location of outsourcing suppliers and sub-contractors

One of the basic questions to be asked here is whether or not the work is to be carried out on site or is manufactured and then fitted on site. Traditionally most components, such as window and door frames and doors were made elsewhere and then fitted on site. However, more recently some of the fitting has been carried out by the supplier. For example, windows arriving on site already glazed, doors already hung in the frames complete with locks and reinforcement steel cut and bent. Historically, there have been periods when considerable amounts of the building have been made elsewhere, including the structure, as occurred during the 1960s when 'industrialised building' was used almost exclusively in the provision of housing. Timber frame construction for housing has recently started to make a comeback after the reaching a peak of approximately 25 per cent in the 1980s. There are significant moves to off-site production as can be seen in the UK Defence Estates single living accommodation modernisation programme (SLAM), using prefabricated modular units.

Whilst the location of the construction site is predetermined, though speculative builders may have some choice, not all companies in the supply chain may be appropriately located for servicing that project. Suppliers may be less of a problem than sub-contractors, as the movement of labour is not involved, the main considerations are the costs of transport to site and environmental implications. There are several factors which should be considered when selecting a sub-contractor in the supply chain or one still not in the chain, besides the criteria that have been used previously. These are:

- Location: Is the sub-contractor based in the area and if not do they have experience in working there?
- Labour: Can the sub-contractor provide enough properly qualified people living in the area or will they have to be brought in for the duration of the work?
- Accommodation: Is there ample accommodation in close proximity to the project if personnel need to stay over during the week?

7.6 Forecasting

Without knowing one has a salary cheque coming in each month it would be imprudent to take out a hire-purchase agreement. Similarly, without knowing the likely workload of the company in the future, it is difficult to negotiate with prospective or current members of the supply chain, unless they can demonstrate there is some guarantee of continuity of work for them.

There are many factors which can influence the potential for obtaining work in the future, as illustrated in Figure 7.5. The ability to forecast future trends has a direct bearing on the capability of the company to prepare for the future and to manage the business to obtain work. This is often referred to as having vision.

The nature of construction means there can be a significant time gap from the client's conception of the idea to the time when the contractor

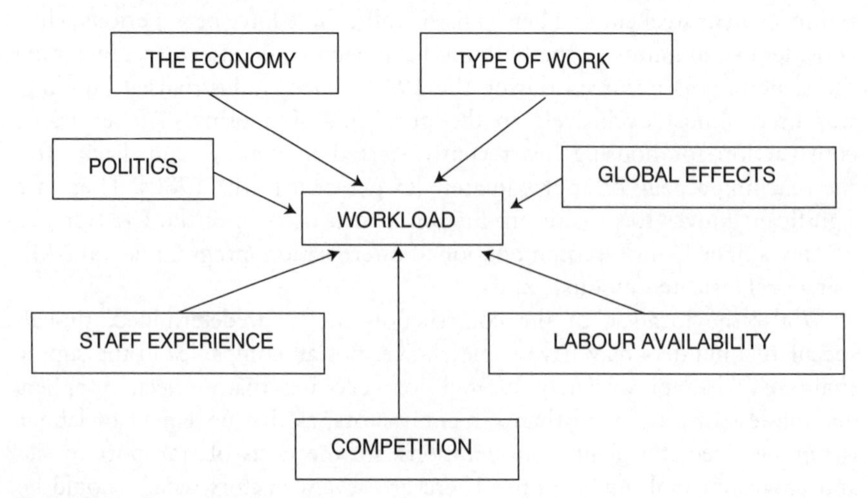

Figure 7.5 Factors affecting future workload

becomes involved, especially in traditional forms of procurement. Unlike manufacturing, contracting is about tendering for work against competition, rather than competing against other products in the marketplace whose success is often being determined by the ability to market products on the assumption the product is well made and the belief it is what the customer requires. Construction output is largely a function of the economic climate determined by government policy and global influences. The type of work in the public sector is primarily linked to the government's political philosophy. Since there is only a finite amount of trained labour, if the predictions are there is to be a boom in construction work, difficulties arise in obtaining labour that can result in imaginative design and construction solutions to overcome these shortages. Accurate forecasting is therefore not easy.

7.7 Managing the supply chain

Supply chain management is not just about managing the supply of materials and labour between the various operations in the chain, but also about managing the flow of information. The information flow should not just be about the current contract but should include information about, for example, the future of the contractor's workload so that those in the chain below are kept fully in the picture. This is a reasonable expectation since their future is to some extent linked to that of the contractor. The supply chain can be considered as a series of activities with storage periods in between. For example as shown in Figure 7.6, material is extracted then stored before being made into a product after which it is stored until built-in to the project where it remains until the building is handed over to the customer.

This is very simplistic as many different materials will often have to be extracted to enable the manufacture of a product, and once in the building, they may immediately start their function, such as a structural element or may not until commissioned, such as a piece of plant. However, by understanding the flow, decisions can be made to manage the chain. For example, investigations of the causes for storage between manufacture and construction, which can take place at the factory and the site, may reveal excessive storage times, increased risk of damage, double handling at both locations, only part loads being dispatched, and restrictions on the sequence of manufacture, all of which contribute to increased costs. This clearly shows that to manage the chain it is essential that all parties understand each other's business requirements and processes, hence the need for an emphasis in building materials education to introduce students to manufacturing processes as well as the properties of materials.

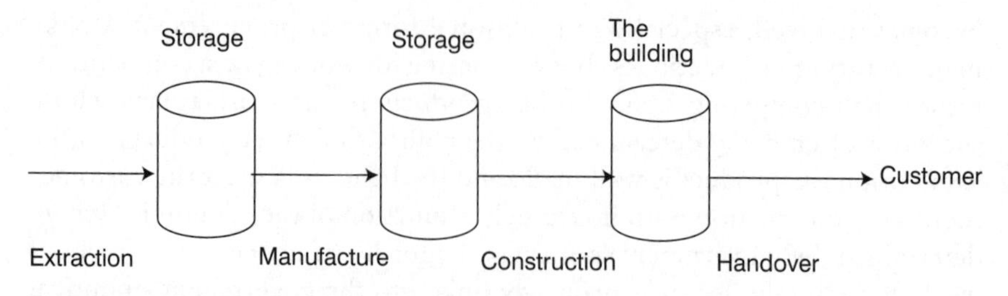

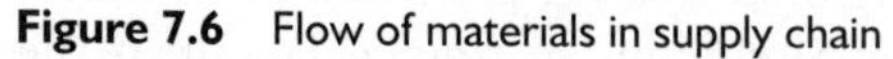
Figure 7.6 Flow of materials in supply chain

Supply chain management in construction is a form of partnering and the same formula for success is required, as laid out in *Finance and Control for Construction,* Chapter 8, that there are clearly defined lines of communication and statement of roles, procedures for resolving disputes, a statement of working in good faith and openness, especially concerned with costs and delivery problems.

After the strategic decision to engage in supply chain management, the procurement of suppliers and sub-contractors follows. Construction purchasing is well established and the significant players in the industry already have in place methods of procuring and selection of preferred suppliers and sub-contractors. They look for key performances against certain criteria such as cost, quality, delivery, working to programme, flexibility and reaction time, experience and reputation.

A dilemma for purchasing managers is to decide if all the work obtained should be given to one supplier or sub-contractor taking up a high proportion of their capacity, known as single-sourcing, or whether the work should be spread about to several, known as multi-sourcing. Both have advantages and disadvantages as shown in Table 7.2, which are similar to the decision whether to outsource or do the work in house as argued in section 7.4.

In contracting there are limitations to the extent of being reliant on one organisation, because sites can be anywhere within the region of work as discussed in the introduction. So the debate is more likely to be about the extent of multi-sourcing that should be considered, especially sub-contracting. Single sourcing for bulk materials is a possibility. For example, a construction company having produced their business plan for the year will have a reasonably accurate prediction of the quantity of reinforced concrete required for the same period and therefore can consider single sourcing the ready-mixed concrete and reinforcement steel.

The whole aim of supply chain management is to improve performance so that all in the chain, from the lowest level to the customer, derive benefit. One of the major difficulties with the chain is the variability of demand and supply. Construction projects are delayed in starting or during the

Table 7.2 Advantages and disadvantages of single and multi-sourcing (adapted from Slack *et al.*, 2004))

Single-sourcing		*Multi-sourcing*	
Advantages	Disadvantages	Advantages	Disadvantages
Improved relationships and understanding Better communication Higher levels of co-operation Higher dependency encourages greater commitment Potentially better quality Possible savings in costs due to guaranteed work Higher confidentiality	Vulnerable to disruption in supply Supplier may have problems in dealing with variable demand Risk supplier can demand higher prices	Competition could drive prices down If one source of supply fails another can be substituted Wider range of expertise and experience available	Long-term commitment more difficult to achieve Quality management more difficult Poorer communications Suppliers less likely to invest incompatible systems Less economy of scale which can affect prices

construction process, suppliers are in turn let down by their suppliers, or they cannot meet the demand asked of them, because other customers have increased their requirements. Any of these can affect the amount of stock suppliers have with direct consequences on their cash flow. The key to this is communication from one party to the other, so that all involved are aware of the issues and can assist in each other in sorting the problem out, or can take alternative steps within their own business. This information should be passed rapidly from one to the other. If a supplier knows that delivery is not going to arrive on site when required, the site needs to know. They won't like it, but it means they can reprogramme their work for the day and reallocate what would otherwise be idle resources. They also have time to mutually discuss how the lost production can be made up the following day.

There are key issues that need to be addressed in achieving effective supply chain management.

Managing communication

The establishment of direct lines of communications between the main contractor and the sub-contractors is essential. This means individuals need to be clearly identified within each organisation and their area of responsibility for decision making specified. This could be one person or several depending on the size of the organisation and the scope of the work. What must not be allowed to happen is a situation where there are 'too many cooks spoiling the broth' or 'the left hand doesn't know what the right hand is doing' resulting in a breakdown in communication. If the contractor is involved early in the process they are more likely to take ownership of the project and will involve the sub-contractors at an earlier stage with the same benefits accruing.

Managing information flow

A problem in construction is the accuracy and completeness of information and the speed at which it is produced. The quantity, variability and fact the drawings to a large extent are unique to the contract adds to the difficulties involved. All of this information needs to be progressed, distributed and controlled to be effective. Any form of partnering necessitates openness in financial matters, and many parties are still hesitant in releasing this information. This is part historical, because of the confrontational nature of the industry and also a cultural in-built puritanical protectionism on matters of money: 'We don't talk about it.'

Dispute procedures

Even though all parties initially enter with high hopes and aspirations, circumstances will arise where disagreements occur so procedures have to be in place to resolve difficulties quickly and effectively. The key requirements are that all parties know the procedures and action is taken immediately the problem arises to establish the causes and agree a solution. The longer there has been a relationship between parties, the less likely there will be disputes. An environment should be created where those in dispute come to the table with a positive attitude expecting a solution to be achieved.

Value engineering

Each organisation in the supply chain has expertise. If this can be pooled in someway, it will inevitably lead to giving the client better value for money.

This is partly a communication issue in getting the interested parties together to discuss and develop solutions, but it is also a timing issue, as to have the greatest impact this must occur at the early stages of the design (*Finance and Control for Construction,* Chapter 7). It is easier to move a column on paper than it is once it has been built. If the parties working in a supply chain are together for a series of contracts then the added value should improve further as all involved more fully understand the client and each other's needs.

Compatible systems

Over a period of time organisations develop systems for running their businesses. These are not necessarily compatible with other organisations they are dealing with. This is especially the case with information technology systems. There needs to be compatibility so the systems can 'talk to each other'. Examples of compatibility are in programming, scheduling, payments, computer aided design, the bills of quantities and quality systems.

Quality

The aim is to continually improve the quality of the completed building. To do this requires a Total Quality Management system that encompasses all parties. All parties should be aware of what the client requires and each other's needs, and the client in turn should appreciate what is achievable within the budget and programme. As with value engineering, the best place to start is to engage interested parties as early in the process as possible so there is more integrated thinking at the design stage. Not all sub-contractors and suppliers need to be or indeed can be involved at this stage, so they need to be engaged during the construction phase by means of discussions and brainstorming the issues. All should be aware of the requirements before they commence work and if problems occur between trades during the work, these should be sorted out quickly.

Long-term relationships

There are some long-term arrangements between contractors and sub-contractors, but these tend to be informal often as a result of the sub-contractors and suppliers being used on a regular basis. There appears to be little enthusiasm at the time of writing to formalise these relationships. There are more formal relationships between client and main contractors, which seem to profit both parties. These are often in the retail and supermarket businesses where the client has a long-term policy of expansion or refurbishment and can see the advantages of having this type of relationship with a contractor. Petrol station contracts have also been procured with long-term contractual relationships.

All parties should understand the importance of relationships, between the client and the contractor, the contractor and the sub-contractors and relationships between departments within the organisation as well as within individual departments. This should not be just about 'getting along', but should be focused with the intent of improving performance and adding value to the processes. This value improvement can be achieved by considering the removal of all processes in the chain that do not add value, addressing all aspects of the costs involved in the chain, openly assessing each others performance and seeking to improve each others capabilities.

Customer relationships are not just between the client and the contractor. The customers are also the users (both employees and visitors) There should be an understanding of all their needs and should be enlarged to include the neighbours in the vicinity during construction and after handover. In the former case, because of pollution and extra traffic, and in the latter case, because the local environment has been altered due to the presence of the new building and its users. Systems should be in place so that all these stakeholders' satisfaction, or otherwise, can be measured so and remedial actions taken if required.

Suppliers are often selected purely on price and previous performance without ever asking what can be done to help them. What do they need to make their businesses more effective? Does the contractor compliment their good performances or concentrate on their failures? When the main contractor requires changes to be made is there a full understanding of the impact this might have on the suppliers and sub-contractors?

References

Bozarth, C. and Handfield, R. (2005) *Introduction to Operations and Supply Chain Management*. Prentice-Hall.

Briscoe, G.H., Dainty, A.R.J., Millet, S.J. and Neale H.N .(2004) Client-led strategies for construction supply chain improvement. *Construction Management and Economics,* 22(2): 193–201

Egan, J. (1998) *Rethinking Construction*. HMSO.

McGeorge, D. and Palmer, A. (2002) *Construction Management New Directions*, 2nd edn. Blackwell.

Slack, N., Chambers, S. and Johnston, R. (2004) *Operations Management*, 4th edn. Prentice-Hall.

Womack, J.P., Jones, D.T. and Roos, D. (1997) *The Machine that Changed the World*, Macmillan.

CHAPTER

8 Quality management

8.1 Introduction

Quality has always been an issue for mankind, be it the area of agriculture and food production, the development of tools for hunting and gathering, manufacturing or health. Japanese manufacturers realised that early attempts at quality control were limited and pioneered quality assurance methods and, more recently, total quality management systems. This may explain why their manufacturing base has been so successful relative to the West, which is now playing catch up. The construction industry is primarily concerned with one-off projects and historically, due to having the design and construction processes divorced from each other, has been left behind. However with the advent of new procurement methods, such as design and build, management contracting, construction management and partnering, the industry is rapidly accepting this change and increasingly adopting quality management systems.

8.2 Quality control

This was the mechanism traditionally used by organisations concerned with quality. Its basis is that of measurement and inspection. In other words, it occurs after the fact as a checking procedure so action can be taken to improve future activities, rather than having systems in place before the product or service is made or offered. The traditional assessment of a quality acceptance level was defined by Juran (1979) as 'fitness for use' and by Crosby (1979) as 'conformance to specification'.

A typical quality-controlled workplace required inspection and measurement points to be established. A method or procedure for the inspection and measurement would be established, after which data would be collected and analysed so that trends in deterioration of quality could be spotted and extreme problems avoided. For example, in the manufacture of precast concrete components, the inspection points would be checking the quality of the raw materials on arrival, the concrete prior to discharge into the mould using standard concrete testing methods such as slump and cubes, followed by Schmidt hammer tests to check the concrete was of a sufficient strength to permit de-moulding. The completed components would be dimensionally measured and had to fall between a range of sizes to become acceptable for dispatch. These measurements would be recorded over a period of time to plot trends, so that as the moulds slackened with use, increasing the size of the component, action could be taken before the situation deteriorated beyond the specified tolerances.

In other words, measurements were made to compare against a specified standard. This meant paying for an inspector, the process added no value to the process or product, reject stock had to be paid for, and, in the end, all these inherent costs would be paid for by the customer as part of the supplier's overheads. What it did not do was to improve the quality of the product or service, but just reduced the likelihood of incorrect products leaving the factory.

8.3 Quality assurance and management

So was quality control alone satisfactory? What about whether the service or product was delivered when the customer required and was it at the cost the customer was able or willing to pay? These were not considered quality issues. It became clear that inspection itself is not enough to sort out quality problems. What was required was a measurement of quality that accounts for all of the customer's requirements. This was referred to as quality assurance and more recently, quality management. This focuses on compliance with procedures to ensure product or service quality.

The British Standard for this was BS5750 published in 1979. In 1987 the International Organisation for Standardization based their ISO9000 series on the BS, later adopted by the EC and the UK as BS EN ISO 9001, ISO 9002 and ISO 9003. These were subsequently integrated to become BS EN ISO 9001: 2000. Note that ISO is the Greek word for equal and not an acronym for the International Organisation of Standardization. All of these have been concerned with setting standards for the *system* of quality assurance rather than the product. They set out a framework by which a management system can be implemented so the needs of the customer are fully met. In other words, quality assurance is based on the principle that it is better to prevent quality problems rather than detecting these problems after the product has been made or the service provided.

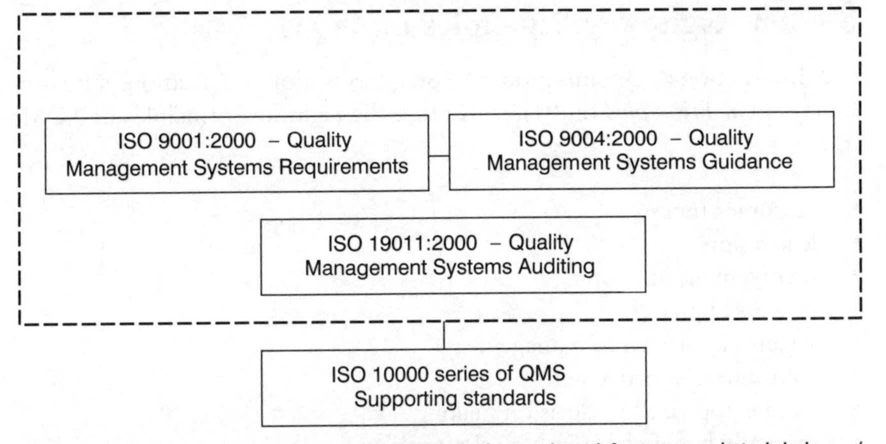

Figure 8.1 The family of ISO 9000 standards (reproduced from www.bsi-global.com/HigherEducation/Quality+Management/IntroISO9000.xalter)

Figure 8.1 shows the full ISO 9000 family as produced by the British Standards Institution although this text will only look in more detail at 9001.

The ISO 9000 family is concerned with the way an organisation's quality management meets the customer's quality requirements and any applicable regulatory requirement, with a view to enhance customer satisfaction and to continually improve the companies performance in pursuit of these objectives.

Quality assurance activities used to be defined to include:

- how the organisation develops policy in respect to quality;
- the allocation of responsibilities for quality within the organisation;
- procedures used to carry out the needs of the business;
- the standards to be attained in the workplace;
- the documentation required to demonstrate both the operation and maintenance of the system and the attainment of quality.

Whilst this still is part of the process, it is very much system orientated rather than people focused. As can be seen in the eight key principles of quality management (section 8.4) referred to in ISO 9001:2000 and ISO 9004:2000, it has become more a management of quality issues which includes people as described below. This is often referred to as Total Quality Management (TQM) described by the Department of Trade and Industry (2005) as:

> A business philosophy that focuses on quality throughout the organisation. It aims to deliver complete customer satisfaction, benefits to all staff and benefits to society as a whole.

8.4 The eight key principles of TQM

The International Organisation for Standardisation publication, *Quality Management Principles* (2001), states that the eight key principles of TQM are:

- customer focus
- leadership
- involvement of people
- process approach
- systems approach to management
- continuous improvement
- factual approach to decision making
- mutually beneficial supplier relationships.

8.4.1 Customer focus

All organisations depend on their customers and therefore should understand their needs, be able to meet their requirements and try to provide a product or service that is better than expected. In reaching this goal, the organisation will find they have become more flexible and speedier in reacting to the market, their own organisation will be better prepared in meeting the customer's needs and this will result in more business as a result of customer loyalty. Historically, in construction, the industry has not been customer friendly, resorting to a more adversarial approach adhering strictly to the conditions laid down in the contract documents at the expense of everything else. However, in recent years this attitude has changed considerably with the development of new forms of procurement such as partnering and management contracting (*Finance and Control for Construction*, Chapter 8).

An organisation focusing on the customer has to research and understand their needs and expectations and ensure this is communicated throughout the organisation. Performance has to be monitored and deficiencies addressed rapidly. This should not be carried out in an ad hoc basis but systematically managed to make it work. However, satisfying customers must be balanced with other interests such as the shareholders, employees, suppliers, financiers, society and the environment.

8.4.2 Leadership

Without leadership it is unlikely the whole organisation will take the matter seriously. Management has to create an environment that encourages all to become fully involved in meeting the organisation's objectives. Good leadership will ensure the employees understand and are motivated to achieve the objectives set. It also reduces the likelihood of miscommunication within the business.

The advantage to the business is a clear vision of the future built to enhance motivation. People know what is expected of them, have clearly defined targets to meet and their successes recognised. Levels of trust increase and fear diminishes as fairness prevails. Staff development is improved and encouraged and their needs increasingly satisfied.

8.4.3 Involvement of people

As a result of good leadership, employees at all levels of the organisation are enabled to use their full range of abilities for the business's benefit especially when the attitude is towards a more participative form of management. Involved personnel tend to be more motivated and committed and contribute ideas to the process, are prepared to be accountable for their actions as they wish to continually improve.

This means they understand the importance of their contribution and bring to the attention of management any constraints affecting their performance. They accept their responsibility and monitor their own performance against their own goals. They look for opportunities to widen their ability and experience and are prepared to discuss problems sharing their ideas and knowledge with others.

8.4.4 Process approach

The ultimate objective is achieved more efficiently if the activities and resources are managed as a process. The benefits include lower costs and speedier operations, more predictable outcomes in terms of quality and time, leading to more accurate planning, and problems are readily identified, rectified or improved.

By systematically defining what has to be done to achieve the desired result, clear responsibilities and accountabilities are identified. Overlaps and gaps in responsibility between the different functions in the organisation are spotted and their ability to manage the situation clarified. It means by looking in detail at the methods of work, how it is resourced and use of materials, the process will be improved.

Finally, through risk assessments, the impact and consequences of the work on customers, suppliers, employees and others can be evaluated and action taken to improve.

8.4.5 Systems approach to management

There are various processes taking place at any one time within an organisation each of which can have an impact on another, such as all the sub-contractors working inside after the building has been made watertight. The aim is to identify all these processes, to understand them and to manage the relationships between them to achieve an effective and efficient solution. In doing this all parties will have increased confidence in the overall management, key processes will be identified and proper integration will take place.

The result will be a more structured approach to integration and the parties will have a common objective rather than working in isolation and blaming each other. There will be greater co-operation with all parties aiming to meet the individually set targets on time and to the required quality standard. Each party will feed information back into the system where they can see problems likely to occur before they actually happen.

8.4.6 Continuous improvement

There is no point in any organisation setting targets and sitting back satisfied when they are achieved. As these are being attained, then higher and improved standards should be set so continuous improvement occurs. There are commercial and competitive advantages of doing this, not just by being cheaper and producing higher quality goods and services, but having the ability to react quickly to opportunities with a more motivated workforce.

To achieve this requires continual improvement on all facets of the organisation, including appropriate training for employees so they can improve and to look for ways to improve the processes and systems. This requires monitoring to track, record, recognise and acknowledge improvements.

8.4.7 Factual approach to decision making

Informed decisions should only be made after an analysis of the available data and information. The result of any decision should be monitored so the business can learn from the mistakes made from incorrect or bad decisions. The organisation should have the freedom to review and challenge decisions and there should be a management culture that accepts their views may not always be right and they should be prepared to change their stance if a better argument prevails. To enable this to happen, there neeeds to be a source of reliable and up-to-date data which is available to those that need it. Any analysis should be carried out using valid methods and decisions should be made based on factual analysis taking experience and intuition into account.

8.4.8 Mutually beneficial supplier relationships

The organisation, and its suppliers, and in the case of construction, the design team are independent of each other. However, there are mutual benefits in working together as it will lead to all parties creating increased value, optimisation of costs and resources through increased flexibility in reacting to situations rather than being tied to traditional contractual relationships.

This means having improved communications, working together with common objectives, sharing information and expertise, and in the case of suppliers in particular, discussing future plans. Based on this, partnering agreements can be developed and the organisation can assist the supplier to improve its processes and systems (Chapter 7).

8.5 An alternative view of TQM

Another way of describing TQM is shown in Figure 8.2, which gives an overall summary of the key issues relating to TQM in diagrammatic form with some different headings, notably survival. The reason for carrying out TQM is to provide customer satisfaction without which the business will lose customer support resulting in decline and eventually ceasing to exist, as many well-known companies have experienced.

In summary, some organisations see validation as something that should be done because others in their field are doing the same and therefore do not to wish to be left out. Those with that attitude will not derive benefit, but will find it an expensive and costly process. Validation has to be conducted in the full knowledge that not only is there much work to be done in obtaining

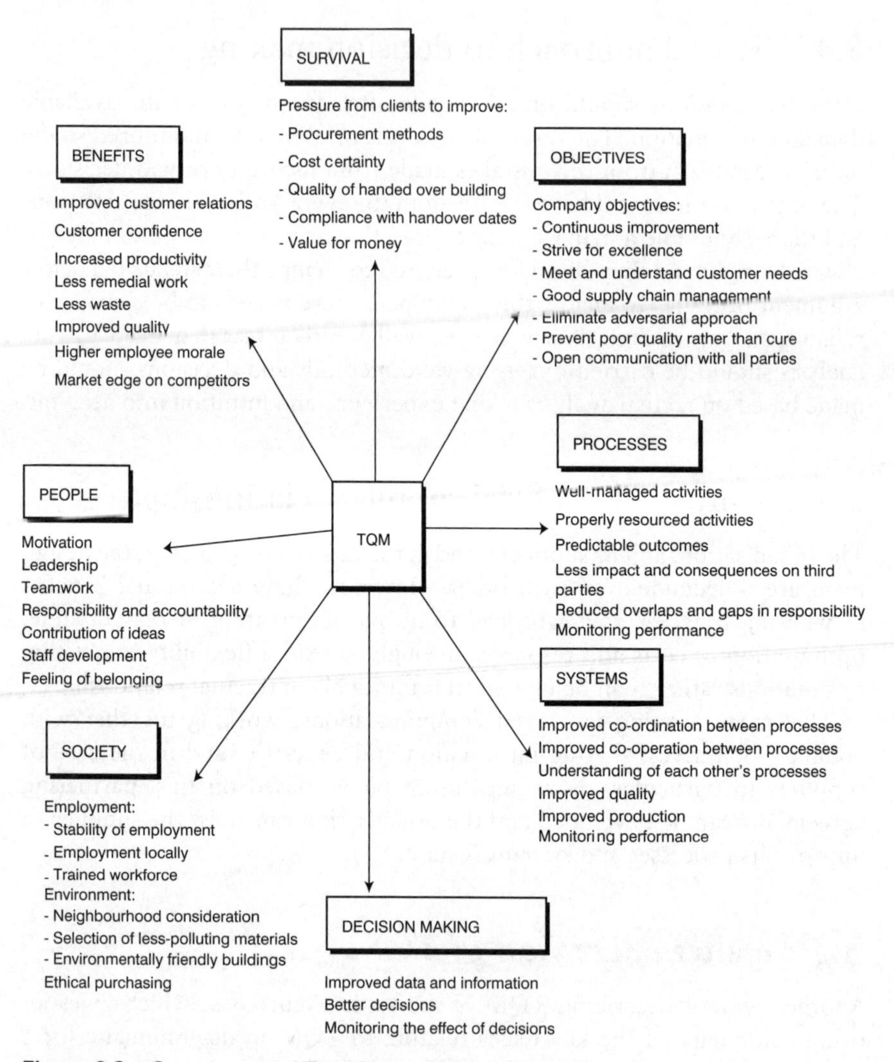

Figure 8.2 Components of Total Quality Management

validation, but that it may well mean some significant cultural changes within their organisation.

As the standard is international and, increasingly, companies of all commercial interests throughout the developed world, especially Europe, are adopting ISO 9000 in lieu of their own national standards, harmonisation is occurring. Customers are also increasingly expecting construction companies to be accredited with the ISO certification.

8.6 ISO 9001:2000 and 9004:2000, the process and implication to the construction industry

To be validated, the organisation has to be audited by an independent company accredited to verify that the organisation is doing as it says and assessing whether what it says is good enough. The audit is

> A systematic and independent examination to determine whether quality activities and related results comply with planned arrangements and whether these arrangements are implemented effectively and are suitable to achieve objectives. (ISO 9001:2000)

The benefits of ISO 9000 implemented properly are:

- improved customer focus;
- increased confidence in the client that the contractor will meet its requirements;
- facilitation of continuous improvement;
- establishment of consistent work practices and quality throughout the organisation;
- improved working relationships with suppliers, sub-contractors and design organisations;
- improved management decision making;
- embedding of positive attitudes to quality in all employees;
- reducing the dependence on individuals so that when they are absent, leave or are promoted there is not a void;
- there is plenty of evidence to demonstrate that it adds value to the business.

The main clauses in BS EN ISO 9001:2000 are summarised in Table 8.1 and are expanded in the following sections in which the clauses taken from ISO 9001 are shown in italics. The author wishes to acknowledge that the basis for much of the comment in this next section is based upon Hoyle (2001) and would be further enhanced if the reader reads this in conjunction with the ISO 9001:2001.

The aim of this section is not to show the reader how to put in place an ISO 9000 system, but to indicate the kind of issues an organisation in construction needs to consider when doing so. Note that although the ISO looks complicated, many of the requirements are already satisfied by standard working practices so it is not so immense an issue that at first it appears.

Table 8.1 The main clauses in BS EN ISO 9001:2000

4. Quality management systems	**6. Resource management**
4.1 General requirements	6.1 Provision of resources
4.2 Document requirements	6.2 Human resources
Quality manual	6.3 Facilities
Control of documents	6.4 Work environment
Control of quality records	**7. Product realisation**
5. Management responsibility	7.1 Planning of product realisation
5.1 Management commitment	7.2 Customer related processes
5.2 Customer focus	7.3 Design and/or development
5.3 Quality policy	7.4 Purchasing
5.4 Planning	7.5 Product and service provision
5.5 Responsibility, authority and communication	7.6 Control of measuring and monitoring device
Responsibility and authority	**8. Measurement analysis and improvement**
Management representative	8.1 General
Internal communications	8.2 Measurement and monitoring
Management review	8.3 Control of non-conformity
	8.4 Analysis of data
	8.5 Improvement

8.6.1 Quality Management systems: Section 4 (ISO 9001)

Clause 4.1: General requirements

This identifies and gives a general description of the processes involved in establishing, implementing and maintaining a quality management system, rather than the procedural documentation. The significant issue is it should demonstrate the system is not just an add-on that has to be complied with, but rather an integrated approach to quality throughout the organisation. For example, has the workforce been involved and prepared for the implications of the system of quality management or are they just obeying instructions and working to a set of rules.

Clause 4.1a

It is not unusual to find each process in an organisation well organised and carrying out its function well, but there might be conflict of interests between the processes at the expense of the overall goals in satisfying the customer's requirements, so the processes and their application throughout the business need to be identified. In construction this failure is often demonstrated in the relationships between members of the design team, the designers and the contractors, the contractors and sub-contractors and between sub-contractors.

Clause 4.1b

It might appear obvious that the sequence of the processes and how they interact should be determined but on closer analysis it is not. Traditionally, a significant proportion of design was completed before the contractor began work on site. In recent years this has changed, as the pressure from the customer to obtain the completed building faster has increased, resulting in different forms of procurement. Similarly the interaction and sequence of sub-contractors and suppliers to the overall production process, has to satisfy the master programme, which means all involved have to be conversant with the overall picture and requirements.

Clause 4.1c

To operate and control the processes effectively it is necessary to decide the criteria and methods. This necessitates asking some very fundamental questions concerning the factors required to achieve the determined objectives. What is necessary to start the process, what is needed to run it successfully and what is required when it is completed? To understand the implications more fully, apply these three questions in relation to a sub-contractor's work on a contract and consider not only what the answers are, but how they will translated into practice, by whom and by what means.

Clause 4.1d

In construction a lack of information is not unknown, due to drawings either arriving late, being incomplete or incorrect, so systems have to be put in place to stop this from happening. Similarly, materials have to be controlled to arrive as required, in the correct qualities and qualities needed to complete the construction activity. With the nature of the business, the availability of the appropriate skilled personnel and supervision is also an issue. Besides the monitoring issues raises in clause 4.1e, equipment such as computing and software has to be available for monitoring performance. Needless to say there can be a cost implication here.

Clause 4.1e

To monitor effectively, employees need to understand what the process objectives are and the ways they are to be measured. Monitoring is a continuing process looking for unusual or unexpected variations in performance or quality. Examples of this would be mixing mortar where the operative is continually checking the water content of the mix, or after staff have returned from a development programme, establishing if they have been able to apply their newly acquired skills or knowledge. Measurement in the context of this clause is about measuring whether the result of the process has met its set objectives in terms of customer satisfaction, quality, completion on time, cost, waste and interaction with other processes.

The analysis is for both the design and production processes. In the case of design this can include ensuring the designers' solutions can be built, the solution satisfies the customer (including the users) and others involved in the production process, identifying pressure points in the process, satisfying regulations, and determining who does what and when. The contractor has to collect data from the monitoring process so that it can be sorted, analysed and presented in an understandable form with a view to establishing the causes and effect of any variation from the plan so remedial actions can be evaluated.

Clause 4.1f

As the targets set are being reached the time comes when the organisation should be looking at improving upon the targets and setting higher standards. There must be realism in this and an understanding that everything does not necessarily have to be improved if the effort involved does not merit the gains made.

Clause 4.2: Documentation requirements

Clause 4.2.2: Quality manual

The quality manual is to show how the management system in place demonstrates how all the processes are interconnected and how they are able to achieve the stated outputs. It should show how the system has been designed, the links between processes, and who does what. It should be able to act as an aid to training new or transferred employees and show how the company complies with external regulations. It should act as the means to analyse possible improvements and act as a mechanism of communicating the company vision, policies, objectives and values throughout the organisation.

The content of a typical manual could include (adapted from Hoyle (2001)):

1. An introduction showing the purpose, scope and application of the manual along with any definitions that may help a reader not necessarily familiar with the jargon used in the business.
2. A business overview indicating the nature of the organisation, its activities, products and services provided. An explanation of how it interacts with other interested parties such as employees, customers (the clients), suppliers, sub-contractors, the design team members, shareholders, trade unions, planners and regulators. The organisation's vision and values, such as ethical purchasing policies, environmental concerns, and safety and welfare of all employed on sites and in the offices.
3. An organisation chart and a description of the main areas of activity within it sometimes referred to as functions, and where these are located.
4. A detailed description of the core business processes such as procurement of suppliers and sub-contractors, marketing, tendering, business planning, human resource management and site organisation. These would be subdivided into control procedures, operating procedures, standards and guides. It should indicate how the key processes are connected, both vertically and horizontally.
5. The relationships of the various functions to the processes.
6. The relationships of locations to processes, bearing in mind the majority of the income is derived from the construction sites.
7. A demonstration on how the organisation complies with ISO 9001, ISO 14001, health and safety regulations, building regulations, etc.

Clause 4.2.3: Control of documents

The term documents includes both paper and electronic media. Control includes every stage of the document's life from development, through approval, issuing, amending, distribution, receiving, usage, storing, security, and when superseded or obsolete, to the final disposal. Security involves issues concerned with fire, theft, computer viruses, unauthorised changes, etc. The documents include policy documents, instructions, drawings, contractual documents, specifications, plans, tender documents, supplier and sub-contractor orders and control, reports and records.

The reasons for the controls are to ensure the documentation is necessary, serves a useful purpose, is correct, up-to-date, goes only to those who need it, and is understandable and comprehensive enough to allow personnel to carry out their job. Only those authorised should see information that is confidential and any documentation required for potential litigation, opportunities in the future and problem solving should be retained.

Clause 4.2.4: Control of records

The records of the organisation are normally generated from a variety of functions or processes usually to satisfy their specific need. Yet they are all the property of the organisation and at a later stage, others not involved in origin of the documentation, may need access to it. This means that a holistic view needs to be taken in the control of the organisations records and individual's sections need to comply with this requirement, otherwise effective identification and storage essential for retrieval is made more difficult. (Note that filing is easy, it is retrieval that is difficult.) Often in construction, amendments to documents are made manually and when made, must be legible and a medium used that will withstand the length of time the documents are to be stored. With electronic scanning, there can be a loss of quality and this should be accounted for. Since documentation takes up a lot of space the length of time it has to be retained needs to be decided. Sometimes this is determined by statutory and legal requirements. If the amount of material becomes excessive, alternative approaches such as microfilm can be considered. The method of storage needs to be secured against tampering and theft, and certain documentation may have to be protected against extremes such as fire and flooding. On disposal of documentation, a system needs to be in place so sensitive and confidential data is dealt with correctly.

8.6.2 Management Responsibility – Section 5

Clause 5.1: Management commitment

Top management cannot delegate their responsibility lower down the chain of command. Management stressing the need to meet customer requirements can cause confusion if at the same time they are emphasising the need to make profit. Whilst the business has to make profit to stay in business, it may well not if customer needs are not met. They are mutually dependent, therefore, there is no need to stress the profit motivation. This may require a change in attitude within the organisation by top management.

A quality policy should be all-embracing and demonstrate the overall philosophy of the organisation. Safety policy, environmental policy, Investors in People and so on, stand alone but they should be part of the quality policy as each is only a measurement of quality in its own area. Thus, a total quality policy is not far removed from the corporate policy. If the policy is just a piece of paper to show compliance, it is not worth the paper it is written on. It must have the full backing of top management and come with a demonstration of their commitment. It should cover all the interested parties in the process similar to the eight principles outlined in section 8.4.

Once the policy is decided, the quality objectives can be derived. However, reviewing the performance against these objectives is often problematic, because it can be seen as a checking procedure, rather than an analytical process to discover successes against which improvement can be considered, and disappointments, so causes can be identified and rectified.

At first glance the provision of resources conjures up labour, plant and material, but it goes deeper. It includes for sufficient time to be made available, the right level of skill and/or knowledge is on hand, sufficient space to work in, access to appropriate information, and enough finance to pay for it all.

Clause 5.2: Customer focus

Constructors traditionally tendered for a contract, built it, put in for any claim they were entitled to under the terms of the contract, handed it over and got paid for it, all without considering the customer's real needs. The fact there was legitimacy to obtain an extension to contract causing the customer problems rarely ever entered the equation. It was assumed the design team, primarily the architect, had addressed the customer's needs.

The customer is now seen not just as the person(s) commissioning the building but in many cases the users as well. For example, the customer may be a hospital trust, but the customers also include the nurses, doctors, technicians, patients and visitors. Further, the needs of an accident and emergency department are not the same as a maternity ward. These needs have to be determined. Equally the ability of the organisation to meet the needs of the customer or potential customers has to be ascertained. By going through this process opportunities can be found, in part by understanding the customer's unfulfilled demands, and in satisfying these, creating more work in the future. This process can also identify the weaknesses and problems in one's own organisation in meeting these requirements that can be identified and rectified.

Clause 5.3: Quality policy

It is suggested that the eight principles of quality management (section 8.4) should act as the basis for the policy. The written policy should be to the point, with positive statements such as 'we will' rather than 'shall'. For example, 'We will develop partnership arrangements with our sub-contractors and work together to improve all our performances'.

The commitment is to all the interested parties. Regulations, legislation and the articles of the company cover some of these. The rest have to be clearly identified and range from the detailed needs of the company to neighbours and the pedestrians passing the construction site. A demonstratively monitored system has to be put in place to ensure these requirements are met.

No policy will work unless it is communicated to all employees so they understand their roles and responsibilities and are allowed to participate in the decision-making process. This is important as they, more than anybody else, understand what the problems are and if engaged can make a profound difference. There may need to be some form of training introduced, not just for their particular part of the process, but so they can see the overall picture and understand where their piece fits into the complete strategy. This means that they understand it affects everybody from the top, down.

No organisation remains static for long as the marketplace changes, new technologies become available, legislation is enacted and the contractor adapts the organisation to meet the new requirements by diversifying, contracting or expanding, therefore, the quality management system must reflect this and can be achieved by having a continuous feedback loop with the employees and at a strategic level, to ensure the quality mechanisms in place are still appropriate for the needs of the organisation.

Clause 5.4: Planning

Clause 5.4.1: Quality objectives

In construction the word 'product' as used in the ISO is alien, so it should be read as the completed design or the completed construction project.

The levels at which quality objectives are set needs to be determined and established by top management at all levels or functions within the business. These would normally be: at the corporate level where the overall objectives for the business are set; at process level, i.e. the process of designing and

Table 8.2 SMART objectives

S	Specific	Objectives should specify what you want to achieve.
M	Measurable	You should be able to measure whether the objectives are being met or not.
A	Achievable	Are the objectives set achievable and attainable within the timescale set?
R	Realistic	Are the objectives realistic with the resources available?
T	Time	By when do the objectives have to be achieved?

building; at the product and function level in that the design and completed building are as the client required and is fit for purpose (note the latter two are not always the same, but should be); at departmental or function level to ensure, for example, budgets targets are met, staff levels are maintained, morale is high and communication systems work internally and externally; and finally at the personal level to ensure the development of skills, knowledge and ability are achieved so as to be able to carry out the other objectives set. Not only do employees have to achieve the levels set, but they should be accomplished within a prescribed period of time. There is a well-tried technique for testing the effectiveness of objectives under the acronym SMART as shown in Table 8.2.

Clause 5.4.2: Quality management system planning

This can only occur once attainable and realistic objectives have been determined for each level as in clause 5.4.1. Since a timescale has been set for each objective, then as with any planning, the methods and resources required can be decided.

From time to time it is necessary to change the systems because of new regulations, changes in culture or to the marketplace and so on. These changes have to be planned taking note of the interfaces between the various processes, so the integrity of the whole system remains intact. The revised documentation should be introduced at the same time and there should be an overlap between the new and existing systems of quality management until the new system has been proven effective. This requires careful monitoring before, during and after the changeover.

Clause 5.5: Responsibility, authority and communications

Clause 5.5.1: Responsibility and authority

This should be normal management practice, but it is still surprising how often these issues are not clearly defined, communicated within the organisation or kept up to date. Defining responsibilities is normally achieved by producing job descriptions and terms of reference, but these do not always specify the limits of the individual's authority. Alternatively, these are so strictly defined that initiative is stifled and the attitude of 'it's not my responsibility' prevails.

One of the mechanisms of communication is the organisation chart that demonstrates (theoretically) the hierarchy between personnel and links between departments in the organisation. These relationships often change over time and the chart ceases to reflect practice (*Business Organisation for Construction*, Chapter 2). The same occurs with job descriptions and terms of reference. The function of departments changes over time with major reorganisations. Much of this will be supported by procedures defining the tasks that have to be achieved, by when and to what level of performance.

Clause 5.5.2: Management representative

To give the subject the prominence it merits, a person should be appointed at the highest level to demonstrate the organisation's commitment. There are many consultants in the field of quality management systems whose expertise can be sought to assist in this process. In a national contracting organisation, it may be that each regional office has a director responsible for the quality management and there is a director on the main board with overall responsibility.

The person responsible for the design and implementation should be given the authority to establish the processes and to produce the supporting control documentation, as well as manage any changes. The monitoring process means all data about all the objectives throughout the organisation have to be collected and analysed so as to identify both failures and successes, and to find opportunities for continuous improvement to performance.

It is a common theme throughout any quality management system, to remind all involved about the needs of the customer, which when working under severe pressure or on a freezing cold windy day on site, is not always at the forefront of an operative's mind. Hence, there needs to be a continuous process which reminds everybody concerned of this so it becomes part of the culture of the business and of those employed. It may be appropriate to use the sub-contractors to help in this.

Clause 5.5.3: Internal communication

A typical example of poor communication is sending information using email to a set of addressees, when many do not require the information. It is important to ensure the correct information is sent to the right people when they require it and in a form they can understand, without causing confusion or inappropriate action. Often information is sent too early when the recipient is not ready for it, due to lack of knowledge, or because the action date is too far removed, so it just sits in the in-tray. On other occasions it is sent too late not giving sufficient time to act upon it effectively. Some information is confidential or sensitive requiring careful thought when considering distribution, whilst others has to be stored for varying periods of time for contractual or legal reasons, before being destroyed. The speed it has to be transferred and the location of the recipient are further factors to be taken into account. These issues have to be investigated before the method of communicating can be determined.

Clause 5.6: Management review

Clause 5.6.1

The quality management system in place needs to be reviewed on a predetermined timescale to ensure it is still working effectively and remains suitable. In other words, is it meeting the requirements in the best possible way, satisfying customers with the most effective use of resources and at the same time taking into account changes in the needs of the customers, society, other interested parties and new or revised legislation and regulations? This is done by regularly reviewing progress or objectives at defined or planned intervals depending on circumstances. At each stage questions should posed such as:

1. Are the objectives being achieved as meant?
2. Are the processes performing effectively as planned?
3. Are the customers satisfied with the company's performance?
4. Are the procedures and policies being used and adhered to?
5. Does the auditing of the system demonstrate it is working effectively and achieving the desired result?

The records of the review should not just be a record like minutes, but should demonstrate how and by whom the review was carried out, the data used, the standards against which the measurement was carried out and the recommendations made. It should also include a SWOT analysis (*Business Organisation for Construction*, Chapter 4).

Clause 5.6.2: Review input

The questions to be answered are whether or not the system is effective and is it appropriate for the immediate future of the business or should it be modified accordingly? This may involve revising the quality policy, setting new objectives and targets, and improving the quality procedures at the interface between different functions and processes.

The audit procedure should not just be about collecting data on the performance of the systems, but should provide information about the attitude of management at all levels to the procedures. It is important to make comparisons with previous audits to note changes and trends in performance. The auditing needs to be designed so that it is neither too detailed, nor misses key issues, and is therefore compatible with what is being measured.

Traditionally, customer feedback in construction comes mainly from the architect or, in the case of civil engineering projects, the resident engineer, but as the industry develops into management contracting, PFI and facilities management, contact with the customer and users of the building becomes the norm. The system needs to check whether the building and services satisfy customer needs, but should also be able to assess their satisfaction. In other words, do the building and services provided meet both the intended design criteria and quality of construction? It may be the initial brief has not worked entirely and this may require the specifications and the design to be modified. Note this can often occur during the construction process.

The collection of data is to compare performance against the set objectives to identify if the same problem is reoccurring so action can be taken to ensure it does not repeat itself. If it occurs frequently and there is no sign of improvement, then on analysis it may be necessary to reorganise or change the method and technique of work, clarify the level of supervision and/or instigate new training programmes. When measuring performance, the actions taken at the previous review should be looked at to see if the corrective or preventive actions taken are working.

There are always changes occurring, as result of external factors or internal factors affecting the organisation. The management systems should have been set up so they can cope with these changes. However in the event of the system not being able to deal with these changes, the system has to be reviewed and modified accordingly.

Clause 5.6.3: Review output

The key issue here is that any decisions made should be so improvements take place as a result, be it the quality of the design and completed building, services provided, or the processes used by the system itself. Any improvement should enhance the design for the benefit and satisfaction of the customer by improving durability and reducing maintenance, and ensuring the specifications targets are met within budget. The improvements can also be in the management of the construction process, such as clearer instructions, safer handling, less waste and higher percentages in recycling. If any further resources are needed, these should be identified.

8.6.3 Resource management Section 6

Clause 6.1: Provision of resources

The resources to be considered are financial, materials, plant and equipment used for construction processes, office equipment and supplies, staff development and, if the contract includes facilities management, building maintenance. The purpose is to establish what recourses are required so the quality management system functions effectively, is continually improved and meets and enhances customer needs.

Clause 6.2: Human resources

Clause 6.2.1: General

All employees involved in affecting the quality of a product or service must be appropriately qualified and competent to carry out those tasks. However, to get to this stage is a much broader issue than just ensuring personnel are up to the task. It involves the total policy of manpower planning and needs to be directly linked to the corporate strategy of the business. If the business wishes to diversify into new areas of work, this will probably mean needing personnel with different skills. An aging workforce will be approaching retirement age and needs to be replaced. Certain businesses have above average labour turnover whereas others rarely lose staff, but occasionally need new blood to bring in new ideas. External forces such as changes in legislation may involve a change in working practices and procedures. In all of these examples, staff will have to be employed or retrained and that requires a system to ensure that the appropriate selection and training takes place.

Clause 6.2.2: Competence, awareness and training

Competence cannot be measured by certificates and qualifications alone, as the employee may not have had an opportunity to practice the skills learned. In work study it is not permitted to study workers until they have gone through the learning curve. Therefore a competent person is one capable of performing the task in hand to the standard specified irrespective of their paper qualification.

To establish the necessary competence, the task to be performed has to be analysed by deciding what has to be achieved, the knowledge needed, the level and type of skills required, the standard of quality set, and the time in which it should be completed. Competences can change as the nature of both internal and external factors alter. For example, a site manager has to be up to date on employment and health and safety legislation or the designer able to use the upgraded computer aided design software package.

The gap between the competences required and the ability of the personnel is what comprises the training needs. How this is bridged depends on the nature of the deficiency. Some can be provided as on the job training by practising the skills, or being mentored and coached by experienced staff. Others require personnel to be sent on specialised courses, and others a combination of both. A similar process is needed when employing new staff whose profile may not match the competence criteria, or when re-deploying staff within the business.

It is easy to send personnel on courses, but how effective are they? This is important to establish not only because this level of competence is required, but if not effective, the cost of the course has been a waste of money. Unless getting involved in a significant and expensive research project, there is no easy solution to measure the effectiveness of the training other than by peer review and observation by supervisors monitoring performance. Some tasks are easier than others to assess, such as basic skills. However, how do you readily monitor the knowledge acquired on a course on, say, waste management? This can only be accomplished if the person can immediately put that knowledge into practice while it is still fresh in their mind so progress and attitude can be observed and assessed by supervisors or assisted by experts in the field.

In all training the trainees must have clearly explained to them the reason for the learning and the expected levels of achievement so they are engaged in the process. They must also know that help and support is on hand when they are either wavering or confused.

Clause 6.3: Infrastructure

Staff who are not housed in appropriate accommodation and without the equipment and support necessary to carry out their tasks will probably fail to meet the customer's needs. It is of particular relevance to construction sites as often the infrastructure provided is sub-standard. Whilst it is difficult to meet the same standards of space and comfort as the main office, the accommodation on site must be fit for purpose and kept orderly and clean. In main offices there should be planned maintenance to ensure the buildings are in good condition and, whereas site offices are temporary, they should be decorated inside and outside. Note the workspace involves the site itself including the buildings or structures being constructed as well as the land within the site boundary. This should also be kept orderly. Maintenance in this context also includes business continuity plans (*Business Organisation for Construction*, Chapter 10) to cover the business if any of it ceases to function.

Equipment on site such as telephones and computers should be compatible with that used elsewhere in the organisation and not older models passed down from head office. It needs to be maintained, and the case of computers, have the technical support necessary to ensure the hardware and the software are working as required. This support will probably not be located on the site, but personnel must be contactable and be able to get to the site in a defined period of time. Plant used on the site is normally hired rather than bought, so planned maintenance is the hirer's concern. However, if owned, then a properly thought-out planned, maintenance programme with clearly defined reactions times in the event of breakdown, must be established.

Clause 6.4: Work environment

The very nature of construction has a profound effect on the environment as many of the materials selected have a high embodied energy, create pollutants at the place of extraction or come from unsustainable sources. The air conditioning, heating and lighting in completed buildings are contributors to the greenhouse effect. During the construction process there can be considerable inconvenience to the neighbours as a result of noise, dust and extra traffic. Construction operatives can be exposed to deleterious materials such as timber preservatives, solvents, dusts and fibres. When the building is completed the occupants potentially can be in contact with deleterious products when off gassing occurs, fibres are released and where there is inadequate clean air and lighting. These issues should be considered primarily at the design and planning of the construction works.

The workplace environment is not just about the physical conditions to which workers are subjected, but also the social and psychological influences. Most people like some form of social interaction when at work and need be motivated to obtain a high degree of job satisfaction. On the physical side, they need appropriate seating if in a sedentary job, the correct range of temperatures, good air quality and lighting, protective clothing, restrooms and so on.

8.6.4 Product realisation Section 7

The term product as used in ISO 9000 is confusing when applied to the construction process as the industry is concerned with the design and construction of a building or an item of infrastructure. Others organisations supply products or components, such as air-conditioning plant, boilers, kitchen furniture, and windows to this process. However, on closer inspection the only significant difference in the process is the manufacture and supply of these 'products or components', so it is suggested the reader substitutes the term construction project for the term product. Construction is classified as a service industry rather than a manufacturing industry and this adds to the confusion in the terminology, as whilst it is true that the industry provides the customer with a service, it also 'manufactures' a tailor-made product.

Clause 7.1: Planning of product realisation

Adapting from Hoyle (2000) the product realisation processes, appropriate to construction are:

- the sales process where the customer's needs are identified;
- designing the project to meet the customer's needs and expectations;
- procuring the materials, components and services needed to accomplish the design;
- constructing the project;
- commissioning the services within the project;
- maintaining the project after handover during the retention period;
- providing technical manuals, information and advice for the successful running of the completed building;
- maintaining and supporting the building in use, known as facilities management.

The sales process in construction varies depending on the type of work being carried out. The speculative housing business has a high marketing profile that relies on the general public as its customers and will spend a considerable amount of its budget on promoting its products. On the other hand, mainstream construction relies more on reputation, competitive tendering and networking to obtain work.

In mainstream construction, since the majority of work is obtained through negotiation and tendering, the planning function occurs at corporate level with the business plan and at each of the different functions, such as estimating, to ensure the tendering can be completed properly, and most importantly, the pre-tender programme, the master plan and detailed plans for the contract (Chapter 2).

One of the major problems in construction is controlling variations. Some of these are unavoidable unless excessive amounts of costly site exploration are carried out, such as problems discovered underground, but many of the others are as a result of the customer changing its mind or failed design solutions. Quality management systems should be designed to reduce these to a minimum and should be part of the overall systems used in the organisation.

The quality objectives for a building can be varied and complex, including making an architectural statement, specifying the standard of finishes, stating dimensional tolerances, laying down standards of 'greenness' and complying with regulations, all within budget and time constraints. Some customers specify criteria for standards of safety, which if not met, will affect any chances of continuity of work. At one level the specifications should determine the standards of performance as with materials, but at another level they should define its purpose or function as with the use of space or a kitchen layout.

Generally each contract has its own set of criteria for quality set by the client and the design team. There are exceptions such as partnerships with customers to build, for example, supermarkets where the criteria will, by and large, be the same for each store. In construction, provision of documentation specific to the contract is commonplace and to a large extent the requirements of this clause are in place. The systems in place to control this documentation, such as drawing registers, should be identical across all contracts the organisation is executing.

Verification of the production and design activities come in a variety of different forms, including having designs and variations checked and signed off, materials and components tested on or off site, visual inspections, comparisons with previously accepted samples of workmanship and the measurement of completed work. The documentation for this should not conflict, so what is stated in the specification is the same as that which is accepted in practice. The frequency of each of these events taking place should also be considered when the customer or the designers have not specified it. Activities that have been specified should have clear criteria as to what is acceptable enough, although these might be challenged if they are set at a different level to that required to suit customer's needs. Where there are no standards laid down, these scenarios should be investigated and criteria for acceptance specified.

Not only are data collected during these processes, but they need to be designed in a user-friendly way and be easily interpreted to determine whether or not objectives are being met. This should include how it will be filed and stored so retrieval is made simple.

Clause 7.2: Customer-related processes

Clause 7.2.1: Determination of requirements related to product

In construction it is standard practice for customers to convey their requirements via the contract documents, which include sequence and handover dates, the drawings, quantities, specifications, confirmation of the price, the time of interim payments, daywork rates, retention durations and retention percentages. Some of these are modified by architects' instructions as the job progresses and any variation to the contract recorded. A similar process occurs with sub-contractors and suppliers and all parties sign up to this as accepted practice in the industry. There are also contractual procedures to claim for extensions to the length of contract caused by certain unforeseen conditions outside of the control of the contractor.

Traditionally, contractors built what they were asked to build in the time period stipulated and on the assumption that information and instructions given were what the customer wanted. However, the customer may have a further set of needs that were not communicated to the contractor. For example, the customer may not fully comprehend how the building will be used in practice. The designers and contractors are in a position to help and give advice on these matters by establishing what the customer actually wants to use the building for. This can more readily be done because of the relationships resulting from the different methods of procurement now commonly in use.

There are a multitude of regulations affecting not only the construction of the building, but also the use. Customers are increasingly taking an interest in the former as any default by the contractor can have an effect on the customer's reputation. It is reasonable to expect that when the building is handed over, the plant and equipment are safe to use and maintain, there are no uncontrolled dangerous emissions and wastes, the interior is safe to the users and materials used in its construction are not deleterious.

The designers and contractors may have limitations imposed from within their own organisation, such as environmental conditions or ethical issues, such as not working for clients directly or indirectly involved with manufacture of arms, declining products made in some countries because of their human rights record, or a requirement to employ a certain percentage of the workforce from the area in which the building is to be constructed.

Clause 7.2.2: Review of requirements related to product

Buildings are one-off projects or, as with supermarket and speculative housing developments, of an existing design modified to suit the site requirements. In speculative work the review is to confirm that the development will satisfy the requirements of prospective customers, which includes the location, design, timing, phased hand-overs, remedial reaction times and assistance offered for finance and legal duties. In the other cases the review is aimed at the processes and procedures at the acceptance stage of the contractual process, from deciding whether or not to tender, through the tendering process to the final decision to sign the contract.

Before taking on any contract the company must be sure it will have the resources and capabilities in place to take on the project. Internally this includes the appropriate expertise, sufficient personnel with time to carry out the work and access to finance to fund the initial negative cash flow. Externally, are the suppliers and sub-contractors capable of carrying out the work as required, and can other specific conditions be met, such as employing a minimum percentage of local labour?

Records of the results of the review and any actions arising from the review have to be maintained (see clause 4.2.4).

Clause 7.2.3: Customer communication

There is inevitably a risk when marketing the company and its services, that a gloss is placed on its image highlighting all the wonderful things it can do and playing down or ignoring its weaker points, but the adage that 'public relations is being good and getting credit for it' should be remembered. The customer should be able to interpret this information in such a way as to be clear what the company can actually deliver and in no way be misled.

There needs to be processes in place that ensure any enquiry is dealt with in a timely manner and the reply contains information that the customer needs to make a decision. Throughout the design and contract period there should be a system in place which ensures the customer is kept informed of the implications of any decisions made by the designers, the contractor, subcontractors, suppliers or the customer.

Not all complaints are justifiable, although in construction the majority probably are. The way these are dealt with should be determined, and depending on where the complaints emanate, how the response is communicated back, by whom and within the prescribed time limit. All complaints should be recorded and periodical reviews made to establish whether there are any trends either in improvement or otherwise

Not all complaints are justifiable, although in construction the majority probably are. The way and by whom they are dealt with should be determined, based on where the complaint emanates, how the response is communicated back all within a prescribed time limit. All complaints should be recorded and periodic reviews made to establish whether there are any trends in improvement or otherwise.

Clause 7.3: Design and development

Clause 7.3.1: The organisation shall plan and control the design and development of the product

In construction, the design process involves establishing the customer's and the ultimate user's needs, carrying out a feasibility study, producing sketch, preliminary and detailed designs, specifications, production drawings, and feeding back any knowledge gained into further work. These stages need to be planned and controlled especially as there is the added complexity of co-ordinating a range of design expertise, namely architects, structural, building service and acoustic engineers, and the design carried out by nominated suppliers. This means at the early stages of the design, the responsibilities and authorities for design decisions should be clearly identified and recorded. These interfaces have to be carefully managed in terms of authority and communication so all involved are aware of the implications of any of the participant's decisions. A simple example would be the architect deciding on a floor to ceiling height without liaising with the building services engineer to establish the amount of space required for the services. An apparently simple error may have significant implications once the construction commences. It is always easier to reposition a column on the drawing board than it is once it has been built.

Design in construction has the further problem that it is not possible to produce and test a prototype, other than small mock-ups of details or sample materials. What is designed and built is what you get. Hence the best time to test the success or failure of the design is when the building is in use. It would be sensible to obtain feedback from the customer so improvements can be made in the future, but whilst this information can be sourced when there is a partnership with the customer and in speculative housing, it is more difficult to obtain for one-off projects.

It is normal for designers to be given a fixed period of time in which to complete the design to a stage when the contractor can commence on site, although this is not always the case. To accomplish this successfully it is necessary to have a plan in place so that progress of the stages of the design can be monitored. The RIBA Plan of Work makes a good vehicle on which to allocate key dates. There are some projects, usually civil engineering projects that are exploring untried technologies and design concepts, where the risk of not meeting targets is more likely. These require even more detailed planning and control involving probability. The design process is only complete when it has satisfied those involved, it meets the customer's needs, is within the budget allocated and can be built within the specified time scale.

Clause 7.3.2: Design and development inputs

Whilst there appears to be similarities with clause 7.2, the difference is this is more detailed. These inputs should be resolved before any design is begun and typically include:

- the purpose of the building and how it is to be used;
- what materials, components, plant, equipment and labour will come to or leave the site during construction and after hand-over;
- specifications of materials and components which reflect their performance in terms of life expectancy, maintainability, environmental impacts and durability;
- regulations that have to be complied with during construction, when completed and in use;
- the timescale and cost of the build;
- ensuring the design is buildable and has been value engineered where possible, even when the budget for the project is generous.

The functional requirements for a building project are often very complex. At first glance it appears this is what the customer requires, but even in small builds such as a house, third-party interests such as access to read the meters, clean windows and empty dustbins have to be considered. On complex projects the third party's needs include deliveries, maintenance issues, visitors such as the customer's clients and customers, and employees. There are performance issues related to the materials and components used in the building in terms of strength, stability, durability, maintenance, flammability, health, environment, modifications, dismantling and disposal, and security.

The designer has to take heed of a raft of statutory requirements and regulations. At one time this was the Building Regulations, but now includes CDM, the implications of COSHH, these relating to plant and equipment for construction issues and many more for when the building is in use, such as anything to do with radioactivity, safety in maintenance, emissions and public use.

Where possible, the designer should use details and solutions previously tried and tested. Why reinvent the wheel? In any building there are a large number of details for which there are well-tried and proven solutions. There is now the capability of many of these to be stored using up-to-date CAD software, however, this should not be allowed to stifle imaginative ideas and experimentation, which would be verified under clause 7.3.3.

Clause 7.3.3: Design and development outputs

During the design process several solutions will be produced especially at the feasibility and design concept stages. Each should be presented in such a way to permit proper comparisons to test which more closely meets the design input criteria, and any variances to those highlighted. It is appreciated the customer can influence this process. Where calculations are involved there needs to be a checking procedure in place.

Clear and detailed specifications and tolerances need to be produced at this stage to ensure materials and components can be procured as required and the contractor knows precisely what is expected when it comes to the build. These specifications should take account of the long-term implications of maintenance for both life expectancy and ease of carrying out the work. Designers should take note of the implications of the legislation such as the CDM Regulations for the construction processes and the Workplace (Health, Safety and Welfare) Regulations 1992 for when the building is in use.

The final designs and drawings need to be formally approved and reviewed before issue for the construction and manufacturing work. It should be noted that in designing a building or structure, there are many stages of the design, which whilst dependent on each other, may have been designed in partial isolation, for example foundations and the structure, or cladding and the structure. These need to be co-ordinated.

Clause 7.3.4: Design and development review

The design process for building varies enormously depending on the size, scale and complexity of the project, thus it is difficult to be prescriptive here. At the beginning of each project, stages at which design reviews are to take place should be identified. Some are already enshrined in the RIBA Plan of Work stages B, C, D, E and F (feasibility, outline proposal, scheme design, detailed design and production drawings), but stages within these processes may also require review.

The reviews are not about ensuring the processes of checking have taken place, but establishing whether the design satisfies the customer's requirements. In other words value engineering (*Finance and Control for Construction*, Chapter 7). In traditional procurement this was normally carried out by the design team, but to comply with this clause, it is necessary to bring others into this process, which can be difficult as many of the key contributors, other than the customer, may not have been engaged. However, on management construction-type contracts, many of the key players such as the management contractor, sub-contractors and suppliers are in place earlier and some of the building's eventual users may also be known. All decisions made should be recorded demonstrating the reasons for the modifications, so progress can be monitored.

Clause 7.3.5: Design and development verification

This clause is unlikely to be relevant to the construction design process as prototypes are unlikely to be used for testing purposes other than with component manufacturers who carry the responsibility for this work. However, where repetition of the same design occurs, such as in speculative housing, then in essence, a prototype is produced and the building can be verified to monitor performance to improve future houses.

Clause 7.3.6: Design and development validation

As with clause 7.3.5 the construction industry only has repeat performances in a limited number of applications, other than speculative housing and some factory shed-type buildings. In these applications a system should be put in place that records customer dissatisfaction and satisfaction so these can be used to improve the performance of future builds. However, even though much of the work carried out in the industry is of a one-off type, many of the design decisions are common so this feedback is useful for future developments.

Clause 7.3.7: Control of design and development changes

The design of a building is a continuous process until the building is completed. It does not finish and then the building is built. This is even more so with the new forms of procurement contract as the design and construction overlap more than with traditional. The construction industry already has a well-established monitoring and control procedure in place, such as the issue of architect's instructions used not just to 'instruct', but also because of the likely cost implications and subsequent claims. To complete the feedback loop, a review of the effect of the change needs to be monitored to ensure the outcome meets the expected performance requirements. It is important though when any change is made, to ensure the effect on other activities does not affect the overall time and cost of the project.

Clause 7.4: Purchasing

Clause 7.4.1: Purchasing process

Although the term product is used in the clause, it can also be taken to mean service provider, such as sub-contractors. Some products or services purchased are for the building itself whereas others such as stationery and office equipment are for the organisation to function as a business. Whilst this clause can be interpreted as purely for those associated with the customer, it seems reasonable the same procedures should be adopted for all purchasing.

There are four main stages in the process the system has to cover:

- the specification process which includes input from the design team and/or the construction team and identifies the quality, quality, delivery dates and duration;
- the evaluation process which identifies appropriate suppliers and sub-contractors, sends out offers to quote, evaluates their response, selects and places an order;
- the monitoring process which looks at the supplier and sub-contractor's performance;
- the process which accepts the suppliers materials or components and, in the case of the sub-contractor, a successful fulfilment of the contract, and settles the invoice.

When a material or component arrives on site it should be fit for purpose and ready to use. The question is, therefore, when controls should be implemented to ensure this objective is met. Equally, with a sub-contractor, are they going to arrive with appropriate resources and if they are also carrying out design, such as the mechanical and electrical services, do they meet the needs of the customer and the contractor? The inference of this is that control may have to be exerted at stages prior to the work commencing on site.

Some supplies and sub-contractors are critical to meeting the overall programme of the contract so these will require more tight controls than those less critical. Consideration should also be given to the level of control depending on the impact on the construction process if the product, material or workmanship is below standard.

A supplier/sub-contractor must be selected only if it has been ascertained they can meet the criteria laid out to complete the contract to agreed levels. The first stage is to select a number of suppliers and sub-contractors who can be approached to quote for the proposed work, often referred to as a preferred list. Typically the criteria against which they could be evaluated would be:

- their ability to deliver on time to the quality specified and, in the case of suppliers, have the capacity to provide the volume required, have a proven track record and are competitive in price;
- have quality assurance systems in place and are committed to continuous improvement;
- meet the health and safety standards set for their industry if a supplier, and to comply with the contractor's requirements when working on site if a sub-contractor;
- provide an after-sales service within a defined time scale;
- have an active and monitored environmental policy;
- are financially sound;
- conduct their business in an ethical manner.

This work can be backed up with interviews when the sub-contractors are invited to make a presentation on why they should be selected as a preferred sub-contractor. This should be carried out in a structured way to ensure each gets equal and fair treatment. In construction it is only occasionally that this would be required of a supplier, but the same rules would apply.

When inviting a supplier or sub-contractor to bid, the correct information needs to be sent to them to ensure they are all pricing for what is actually required. The procurement documentation should be designed with this in mind so that no issue is missed. These will include location of delivery or work, timing and sequence, specifications, quality assurance systems, compliance with health and safety regulations and so on. There needs to be a system in place that checks all submitted bids for accuracy and conformity with the bid request *Finance and Control for Construction*, Chapter 10).

Records should be maintained of these processes. These are expensive processes, so records of those that have failed to meet the criteria are equally as important as those that have succeeded so unnecessary work is not repeated in the future. These records must show the criteria measured, as a future contract may have a different set. The loop should be completed with feedback from actual performance so a supplier or sub-contractor who has under performed is not re-employed.

Clause 7.4.2: Purchasing information

The supplier must know precisely what product or services the purchaser requires. In some cases the product may be from a catalogue, in which case the literature should be inspected and the unique reference and details quoted in the order. Specifications are often stated in the bills of quantities, or similar documents and these should be used when ordering. Sometimes the specification cannot be matched exactly and the nearest available should be used. This would normally be of a higher specification unless it has been agreed with the designers, as it could be that the lower one satisfies their needs.

When purchasing a service a clear statement of needs must be given. Much of this will be included in the standard forms of contract designed to protect both signatories, but further explanation may be required as part of the contract. There may be a requirement that work is carried out only by specified qualified personnel such as chartered members of professional institutions.

Materials and services can be purchased either verbally or with documentation. The latter is well covered normally in that the precise nature of the order is written down and a copy kept to check delivery and accuracy of the order. The verbal order is more difficult to control because it may be a telephone instruction and the detail not written down, relying more on memory. This is quite likely to happen on small works, but even on larger contracts in the case of an emergency, the contractor may ring up the supplier and ask for more material. In all cases the person making the verbal request must have the authority to do so. Further, there must be a system in place that records on hard copy what precisely was ordered.

Clause 7.4.3: Verification of purchased product

It is essential to confirm that what was purchased was delivered. Some products can be inspected at the supplier's premises before dispatch to site, but this is not common in construction. All other materials and products should go through an inspection process as they arrive on site. This may take the form of individual inspection or random selection depending on the nature of the product. One would expect to check precast components as they are off-loaded, but ready-mix concrete can only be sampled. The method of inspection should be clearly identified before materials and products arrive. The person carrying out the inspection needs to be informed precisely what is being checked for and if the product is found to be unacceptable, follow a set procedure to advise the provider, obtain replacements and dispose of the reject.

Clause 7.5: Production and service provision

Clause 7.5.1: Control of production and service provision

The emphasis is to control quality throughout the production process rather than by inspection at the end. Thus each stage of the process or service has to be identified so appropriate controls can be put in place.

The information required in construction is usually in the form of specification, drawings, the bills of quantities and other contract documents. Unlike in manufacturing the information can include external issues such as noise and pollution limitations, which impact the quality of the environment for the neighbourhood as well as the customer's reputation. Since in construction all the information is rarely available at the beginning of the contract, it is important to plan when and how information will be gathered throughout the project. Further, it would be unusual if drawings were not revised during the design process, so it is essential to devise a method of indicating these revisions and ensuring all interested parties are working from the same updated information.

The information referred to in this clause is to do with informing personnel what do rather than informing them how to carry out the work. This means providing work schedules indicating what the tasks are, when they have to be completed and the availability of plant and equipment to accomplish the tasks, the specifications that have to be met and the methodology to be adopted for verification of the work.

Any equipment provided to assist in accomplishing a task should be selected so it is appropriate and properly maintained and records kept. A piece of plant may not always be fully utilised as it may be expected to carry out many different tasks. For example, a tower crane will not always lift to its maximum capacity. However by careful thought and planning, the effective use of plant can be maximised.

Measuring involves surveying the building site to determine the depth of excavations, position formwork and columns, height of floors and ceilings, then on completion of the work checking the work is in the correct place. There is also the measurement of bought-in components to ascertain whether they are within tolerances and measuring and testing the quality and quantity of bulk materials, such as concrete. Operatives should understand the implications of errors made or found.

Post-hand-over (post-delivery) service is covered to a certain extent as part of the contract. In construction, in the event of poor service or the contractor ceasing to trade, the customer is protected to a greater or lesser extent by the retention clause. Whilst this is acceptable as far as it goes, it falls short of the needs of a good quality system. The contractor may for good relationships, decide to extend the work contract and the contractual length of time beyond the retention period. The system needs to consider the time of the response, the lines of communication between customer and contractor, monitor performance and feedback of the work, so modifications to further work and practices can be modified for continual improvement.

Clause 7.5.2: Validation of processes for production and service provision

There are some processes that can be validated by inspection, or simple testing, but others require very sophisticated and expensive methods or even testing to destruction. In construction, welding would be an example where it would be necessary to use X-rays or gamma radiation methods to verify the integrity of the work, in addition to visual inspection. The emphasis in these cases is to ensure qualified personnel are employed and the environment in which the work is carried out is conducive to good workmanship. Qualified personnel can be confirmed by qualifications and a demonstration of training.

Clause 7.5.3: Identification and traceability

Generally speaking, this clause does not apply to construction as it is concerned with ensuring materials that are similar in appearance go to the right place. There are occasions when this might occur as in the case of some timber preservatives, where the outward appearance of the timber is no different to untreated, and in the case of fibrous materials used for fire-stops, sound or thermal insulation. Where confusion can occur methods need to be adopted for identification. In construction colour coding is often the answer. The same technique is used in buildings to identify uses of service pipes where there is a well-established British Standard colour-coding identification.

Traceability is becoming an issue in construction mainly because of environmental issues. How does one know the timber being used has come from sustainable sources? The supplier using a number code from which the source can be identified normally does this, but the reader should be aware this is no guarantee of its authenticity, unless the integrity of the import supplier is known.

Clause 7.5.4: Customer property

The customer may supply materials and components that have to be incorporated into the building by the contractor or sub-contractors. The documentation accompanying the delivery belongs to the contractor as their record of receipt of goods, but should be passed to the customer when the building is handed over. Special care needs to be given to such goods, records kept, appropriate clearly identified storage areas provided and the goods marked so they can be identified as belonging to the customer. The latter is also necessary to distinguish the goods from others included in the monthly valuation.

On receipt of the goods a careful inspection should be carried out to ensure the goods have been received in good condition and are as on the delivery ticket. Any mismatch in this, damage or suspicion of the same, should be notified to the customer immediately.

Clause 7.5.5: Preservation of product

This clause is particularly relevant to the suppliers of materials and components to the site, but it also has implications to the site processes. Once the item has arrived on site it becomes the responsibility of the contractor to maintain it in an undamaged usable state, thus the way it has been previously packaged is of importance. There is a relationship between the two in that the packaging designed for the supplier needs to conform to the needs of contractor. The supplier will have been primarily concerned with packaging for storage at the place of manufacture and for transportation, in conditions that are often in a protected environment, whereas once on site they can be exposed to the outside elements where both product and the packaging can deteriorate if unprotected. The supplier should be made aware of these difficulties and take account of it in their design.

Identification is important if the contents of the delivery are not visible (in construction they often are), but the markings should not deteriorate if exposed to the outside elements for any length of time. If materials have a limited shelf life, this should be clearly seen so that stock can be rotated.

Hazardous materials must be identified on the packaging so the material can be safety handled, stored and the residue disposed of. It may be appropriate to have assembly instructions also shown on the packaging in lieu of an enclosed instruction manual.

Handling issues comprise the method of handling to ensure safe working and protection of the product. The packaging can be used to assist in this such as voids in brick packs for forklift trucks, paletted goods, or identified as 'fragile' or handle 'this way up'.

There needs to be a system in place to control the quality and the issuing of items from the stores. In construction this tends to be used for the smaller items held in a manned secure building, but much of the larger items such as aggregates, sand, brick, blocks, timber, pipes, timber components, reinforcement steel and precast concrete are left in the open. Some of these can be stored near to the place of work to eliminate transport and excess handling, others in a secure compound to reduce the likelihood of theft. A decision should be made as to which materials are tightly controlled when issued and those which are not, and who is authorised to access these areas. Part of the issuer's function should be to check the state of the remaining stock and to ensure when a package is opened to issue part of the contents, it is resealed. Cross-contamination issues need to be addressed such as sands and aggregates used for concrete mixing, and certain materials will have to be stored separately such as flammable materials and explosives.

Clause 7.6: Control of monitoring and measuring devices

The obvious measuring devices used in construction include surveying equipment, torque spanners and weighing equipment on concrete mixers and cement silos, but there are also other measuring devices for safety such as alarms on cranes when they are approaching the limit of their safe lifting capacity. There are other monitoring and measurement issues such as customer satisfaction, but these are described in section 8. The identification of these devices is because without them, the quality of the final product cannot be guaranteed, or in the case of safety, there is an increased likelihood of an accident.

There are several issues to be addressed when measuring. These are what has to be measured, by what means, to what accuracy, how reliable and accurate is the measuring equipment, how qualified is the measurer, and where and when is the measuring to take place, as for example, in setting out columns on site, or off site for concrete cube testing in a laboratory. Laboratories, be they external or set up on a large civil engineering site, should be certified to the standard ISO/IEC 17025.

Measuring and monitoring devices have to be calibrated at prescribed intervals, although certain devices such as thermometers would not be. This will depend on the type of device, the amount and kind of usage. Equipment used in a well-controlled laboratory does not get the same level of misuse as a device being used on site. The amount of variance measured at the calibration, known as drift, is recorded and if the trend on subsequent calibrations increases, the frequency of calibration should be increased.

Once the calibration measurements have been made the device should be cleaned, maintained if necessary and readjusted back to the standard. The device should then be 'tagged' with the date of the calibration and date of the next calibration. If the device is found to be outside the limits set for accuracy, the quality management system should permit for remeasurement of work the device was used on. Records of the calibrations of each device should be kept and maintained

8.6.5 Measurement, analysis and improvement – Section 8

Clause 8.1: General

The characteristics to be monitored and measured have to be identified so the methods adopted for measurement can be determined.

Clause 8.2: Monitoring and measurement

Clause 8.2.1: Customer satisfaction

The use of the term perception rather than satisfaction is interesting here. It may be that the building is sound but the customer's perception is that they are not satisfied. Who the customer is should also be defined. One customer is the client procuring the building, but other customers include the other stakeholders, primarily the users, of the building.

Information can be monitored in a variety of ways, some directly from contact with the customer, and others from other sources. These latter ones include being asked to carry out more work, how successful the competition is in obtaining work from the same customer, and compliments received either directly or indirectly. More formal ways include meetings with stakeholders, structured interviews and monitoring the number of complaints received during the retention period.

Clause 8.2.2: Internal audit

The purpose of the audit is to verify whether the set levels of performance are being achieved using monitoring and measurement. The duration of the planned intervals will depend on the nature of the processes being monitored. Quality objectives in an office environment may not change much over a considerable period of time, so annual audits may not demonstrate any significant variances. However, the many processes on site are much more varied, often of short duration and faster moving, requiring more frequent monitoring and not necessarily at fixed intervals.

The aim of the audits is to verify that the agreed policies have been implemented, the organisation has met its agreed objectives and the product design and construction requirements have been met. Currently any certification body audit cannot be used in lieu of the internal audit although there is a school of thought that believes that, since they are carrying out the same task, this should be permitted. Conformity with the international standard is demonstrated by conducting a planned audit based on status and importance, and comparing against ISO 9001, or by analysing all data from other audits such as in policy, project and product.

Effective implementation establishes if the processes are working as planned, if the outcomes are having the desired effect, if they are effectively maintained, and if the processes are still carrying out their functions in spite of changes that have occurred as the business develops.

As with any management system, there needs to be an overall plan and programme to indicate when the various audits should take place to ensure all that is necessary occurs and the audit procedures are resourced properly. These audits include management audits such as policy and strategy, product audits for construction projects, and conformity audits for the relevant regulations and standards. The scope of the audits is to define what has to be covered. Since this also includes the limits, it is a useful indicator to check if other audits overlap or parts of the business are missed.

It would be impractical in many organisations, especially small ones, to have totally independent auditors. The selection depends on the integrity and competence of the person concerned so they can be impartial and objective in spite of having friendships and working relations (both good or bad) with those being audited. What is forbidden is auditors auditing their own work. A typical auditing procedure should include the following as shown in Table 8.3.

The post-audit report should also indicate activities not audited and identify activities where there are opportunities for improvement. Where opportunities have been identified, the person in charge of this work may

Table 8.3 Auditing procedures

Preparation	Preparing the annual audit programme for the organisation	Planning the individual audits	Selecting and training the auditors	Producing the supporting paperwork and forms for the audit
The audit	Conducting the audit	Recording the observations		
Post audit	Reporting the audit findings	Establishing corrective actions required	Implementing corrective actions	Checking the effectiveness of the corrective actions

not have the authority to make changes as they may have a knock-on effect elsewhere, mean capital expenditure, or changes in staffing levels. In such cases the report should be given to the correct level of management that has the authority. This should be done promptly as any delay may have implications on future results and performance. The speed of any corrective action will depend on the problem identified and could take a few minutes to several months depending on the complexity of the action needed. There need to be follow-up audits to check that corrective action has been taken and to measure the success or otherwise of this action.

Clause 8.2.3: Monitoring and measurement of processes

There are different ways to monitor and measure processes. For example, in construction visual observation is commonly used as a means of inspecting workmanship quality using either an experienced eye or comparing against a constructed and agreed sample of workmanship. Mechanical devices such as scales are used to weigh materials, control charts are used to monitor progress, measurement of the flow of materials to the site and subsequent calculation of waste percentages, and cost control systems to check expenditure. Annual appraisals are used to monitor staff performance and development, and records kept of staff turnover. All of this should be monitored and audited on a regular basis to demonstrate that the required data are available, and that results are being used for performance analysis so that corrective action or improvement can be implemented.

All of this monitoring and measurement is only of use if it can demonstrate the process is performing as planned, in the best way, and is meeting the organisation's objectives. To do this means clear targets have to be set beforehand.

Clause 8.2.4: Monitoring and measurement of product

Verification in this case is carried out to see if the product or service meets the requirements. This process is staged and commences at the design stage as it is, for example, easier to rectify an error on the 'drawing board' than after it has been constructed on site. In construction this verification can be divided into four major stages: the design process, construction phase, when commissioning plant and services, and finally when the building is handed over to the customer, usually referred to as 'snagging'. In a quality management system this last stage should not just correct errors, but check all the previous verification procedures have taken place.

The evidence of conformity will require the contractor to collect and collate all of the data from the suppliers and sub-contractors verification procedures, as well as their own.

When a product is released and passed on, and this includes the suppliers to the contract, the person responsible for this must be identified. For many processes on site this could be the foreman employed by either the contractor or sub-contractor. The customer may require that some or all of these are also signed off by the clerk of works, or members of the design team, such as the architect or structural engineer. At the design stage, it could be the team leader or a partner of the organisation.

The product release approval procedure requires that the product is seen; the conformity requirements of the product, evidence of its conformity and authorisation may also be seen.

Clause 8.3: Control of nonconforming product

A nonconforming product is one that does not conform to approved product requirements, such as the specification, the organisation's requirements, customer expectations, or if damaged. If the product cannot be repaired or reclaimed it should either be recycled or destroyed. The nonconforming product should be identified as such; in the case of drawings a typical example is using a 'superseded' stamp to demonstrate the drawing is out of date. In the case of the construction process defect materials and equipment, if left on site should be marked in some way or placed in a specified storage area sometimes referred to as 'quarantine' areas. In the case of defective workmanship it is not usually necessary to mark the work, as dismantling takes place almost instantly or, if in dispute, clearly identified in the dispute documentation. Clear instructions as to what is to happen to the nonconforming product should be given, specifying if it should be returned to the supplier, the manner of disposal, since there may be safety and environmental issues regarding tipping, and how it should be recycled, repaired or modified.

It may be necessary to obtain permission from the customer to carry out corrective action, especially if the product is a structural element, and a demonstration of the permission provided. There may be specific instructions, for example that repaired honeycombing in a standard precast bridge beam be positioned in a certain part in the bridge where the loading is lower. This would have to be authorised by a qualified person such as the resident engineer and a documentation and verification trail provided to demonstrate that the product now conforms and has been used where specified.

Wherever nonconformities occur there need to be records kept which show how the nonconformity was discovered, what it is and what action was taken, which includes returning to the supplier, disposal, recycling or corrective measures.

A nonconformity may well affect other activities with potential disruption and delays to the programme. This should be investigated and steps taken to eliminate or reduce this impact. If the nonconformity should have been spotted earlier, an investigation as to why it was not should be carried out.

Clause 8.4: Analysis of data

It is essential that the quality management system itself be verified to see if it is functioning effectively and meets the needs of the organisation. To measure effectiveness requires data to be collected and analysed so improvements can be seen, the targets set are being achieved, the process is being done in the most effective manner, and the process objectives are still relevant to the organisation's objectives.

If the quality management systems in place continue to be used as set up originally and have not taken account of the fact the organisation's objectives have changed, then clearly it will fail. Thus the systems need to be evaluated accordingly. This requires the processes to be reviewed irrespective of whether or not objectives have been set. Where identifiable objectives have been set, these should be checked to see if objectives have been met, targets raised with a view to improvement, and at the same time investigate the impact on the processes and to eliminate targets if they no longer are required to attain the organisation's objectives. Where none have been set there is the opportunity to capitalise on this and improve the organisation's performance by investigating what is happening and setting targets and objectives.

The data outlined in clause 8.2.1 should be analysed and conclusions produced so customer satisfaction can be assessed. In construction some of the data are readily available such as repeat orders, but establishing data on complaints and compliments requires systems in place to record and obtain this information. This process of collection should consider where the information is gathered from, the method used, and the frequency. The method of analysis needs to be determined so trends can be established and methods developed to quantify customer satisfaction.

Conformity to product requirements is answering the question, does the building conform or not to the original, or modified brief? Included in this are the data from the design and construction processes and the customer satisfaction data. Whilst many building projects are one-off contracts, the lessons learnt from the data collected can be used to prevent the same mistakes being made again. For example, details that don't perform as expected or are difficult to construct, and waste exceeding the targets set.

Suppliers – in the case of construction, sub-contractors – are separated in the standard. Their performance can be analysed against a variety of criteria based on the likely impact of their failure on the business. This should not be taken as just a building contract issue, but their contribution to the business as a whole, as the same supplier or sub-contractor may be servicing several contracts at the same time. This may be extended to include design teams.

The number of contracts for which suppliers are used and the value of their work as a percentage of the overall workload is important, as if they fail to live up to expectations, the satisfaction of several customers may be put in jeopardy. Therefore more attention should be placed on controlling those who are the major contributors to the organisation. The same applies to a specific contract if they are a high-value contributor.

The quality of any of the sub-contractors or supplied materials and components, can have impact on customer satisfaction even though the cost of remedial action may be small. This is particularly the case on parts of the building immediately visible to the customer, or such items as services or equipment that do not function.

Delay of the delivery of any product or service activity on the critical path needs special attention as delay to the final completion of the building will cause concern to the customer and may result in penalty charges to the organisation, as well as impact on its reputation. Some items have long lead-in times from the placement of the order until delivery. In these cases consideration should be given to building in some latitude, as the longer the lead-time, the more likely there is to be slippage.

Finally, the issue of costs should be investigated. If a supplier or sub-contractor is to be regularly used thereby guaranteeing work for them, it could be possible to set targets for them to reduce their costs. This has to be linked with the former issues of their performance in quality and delivery.

Clause 8.5: Improvement

Improvement is a continual activity which can be enhanced by feedback from customers and other stakeholders and interested parties. Continual improvement is defined in BS EN ISO 9000:2000 Quality Systems – Fundamentals and vocabulary as:

> The aim of continual improvement of a quality management system is to increase the probability of enhancing the satisfaction of customers and other interested parties. Actions for improvements include the following:
>
> a. analysing and evaluating the existing system to identify areas for improvement;
> b. establishing the objectives for improvement;
> c. searching for possible solutions to achieve objectives;
> d. evaluating these solutions and making a selection;
> e. implementing the selected solution;
> f. measuring, verifying, analysing and evaluating results of the implementation to determine that the objectives have been met;
> g. formalising changes.'

In other words, it is a continuous activity which can be enhanced by feedback from customers and other stakeholders and interested parties.

Clause 8.5.2: Corrective action

Corrective action is not just a case of carrying out remedial work to resolve the current difficulty, but about tracking back the problem to establish the cause or causes with a view to preventing the nonconformity from happening again. It is not a case of taking corrective action for every minor incident but rather to concentrate on the main issues identified. It could be that when and if all the major problems are corrected, the minor ones take more prominence in the future.

The collected nonconformity data should be analysed to determine the causes of the nonconformity, to check whether the nonconformity had been predicted as a possibility, and if so, why the preventive measures did not work. They should be listed in terms of the frequency of occurrence so the major and minor incidents can be determined and the cost of correction and/or savings that would be made.

Determining the causes of nonconformity is crucial in this process. It is not enough just to ask the question why something went wrong, but to continue asking why until the root cause is discovered. For example, at the weekly planning meeting reviewing progress, asking 'why was the brickwork not completed in time?' and receiving the response 'because the scaffolding was not ready' only answers the problem in part. 'Why was the scaffolding not ready?' may reveal another problem and so on. If this questioning process is not completed, the corrective action taken as a result of an earlier answer will result in inappropriate action being taken. Where it is not possible to use this approach another way is to list the nonconformities and brainstorm possible causes in turn. There are many more sophisticated techniques for problem solving if required.

Having established the cause, the next problem is to ensure it does not happen again. In some cases it may be decided the likelihood of the same thing occurring is negligible, or if it does, the impact will be minimal so no action is taken. However, if there is a recurrence and the impact is significant, action must be taken. The solution is to see what action needs to be taken if the nonconformity is between these extremes. A way of doing this is a risk assessment. This can be done using a scale of risk from say 1 to 10, where at the extreme, 10, the contract may come to a halt because of nonconformity with building regulations or a structural defect, and 1 could be no effect. Others in between may include 'causes customer dissatisfaction', 'some dissatisfaction', and 'only noticed by discerning customers'.

If action is taken based on the root cause of the nonconformity, it is more likely to be successful. Not getting to the root cause will normally result in a temporary fix only. Some corrective action can be carried out immediately without any impact on other processes, but often there is a knock-on effect elsewhere. When this occurs the impact of the corrective action on these other processes has to be analysed and incorporated into the overall corrective action.

Keeping records of the causes and action taken is necessary for future reference, to review the corrective actions taken, to evaluate whether they were done in the best possible way and to find out whether the nonconformity has been eliminated.

Clause 8.5.3: Preventive action

In construction, designers experiment with new designs, manufacturers create new products and the contractor tries out new methods. They use new sub-contractors and suppliers and decide whether work should be carried out on or off the contract site. Further, the industry modifies or designs new procurement contracts, governments produce new regulations and laws. In other words: the industry is in a continuous state of change. The organisation needs to anticipate the implications of these changes and take preventive measures so corrective action is not required later.

To predict potential nonconformity, the corporate, product, or process objectives have to be determined otherwise, by definition, possible nonconformity cannot be predicted. Personnel then have to be selected to consider the risks of nonconformity. They will either work in teams or as individuals depending on the nature of the work, focusing on the critical factors most likely to impact on a successful achievement of the objectives. Each of these should be analysed by asking questions to establish what might cause a failure in attaining the objectives, and if the failure occurs, what and how severe would the impact be on the organisation and the customer. The probability of the event occurring is important because if the probability is very low it might mean no action would be taken. Finally, determining the likely cause of failure would be crucial in deciding on what preventative action to take.

There are two types of controls, those that detect a variance or occurrence so remedial action can take place and those that highlight a failure and lead to a permanent change. All existing procedures should be examined from time to time to determine whether they are still appropriate to detect failures or indeed necessary as some may have become redundant as the organisation moves on.

The action required to prevent a nonconformity can take many forms, including redesigning part of the building or one or more of the processes, altering the behaviour of personnel, redesigning the working environment and introducing new methods, safety measures or procedures.

Recording the actions taken, besides being a requirement, can have other advantages in that it demonstrates the organisation has been vigilant in looking out for the customer's needs and complying with regulations, and in the event of a failure, demonstrates the organisation has not been negligent if unfortunately it went as far as litigation. To complete the loop, reviewing the results of the preventive action taken confirms or otherwise that the action has been effective, and whether the effort and time spent has been worthwhile.

References

Ashford, J. (1989) *The Management of Quality in Construction*. E&FN Spon.

Bell, D., McBride, P. and Wilson, G. (1994) *Managing Quality*. Butterworth-Heinemann.

Chartered Institute of Building (1989) *Quality Assurance in the Building Process*. CIOB.

Ciampa, D. (1991) *Total Quality: A User's Guide for Implementation*. Addison-Wesley.
Griffith, A. and Watson, P. (2004) *Construction Management – Principles and Practice*. Palgrave Macmillan.
Hoyle, D. (2001) *ISO 9000 Quality Development Handbook*, 4th edn. Butterworth-Heinemann.
International Organization for Standardization (2000) BS EN ISO 9000:2000 Quality management systems – Fundamentals and vocabulary. IOS.
International Organization for Standardization (2000) BS EN ISO 9001:2000 Quality management systems – Requirements. IOS.
International Organization for Standardization (2000) BS EN ISO 9004:2000 Quality management systems – Guidelines for performance improvements. IOS.
International Organization for Standardization (2001) 'Quality Management Principles'. IOS.
Kantner, R. (2000) *The ISO 9000 Answer Book*, 2nd edn. John Wiley & Sons.
Sasaki, N. and Hutchins, D. (1984) *The Japanese Approach to Product Quality*. Pergamon Press.

Index